# The Classification of Quadrilaterals

## A Study of Definition

A volume in
*Research in Mathematics Education*
Barbara J. Dougherty, *Series Editor*

# The Classification of Quadrilaterals

## A Study of Definition

*Prepared by*
**Zalman Usiskin and Jennifer Griffin**

*with the help of*
**Edwin Willmore and David Witonsky**

INFORMATION AGE PUBLISHING, INC.
Charlotte, NC • www.infoagepub.com

**Library of Congress Cataloging-in-Publication Data**

Usiskin, Zalman. The classification of quadrilaterals : a study in
definition / prepared by Zalman Usiskin and Jennifer Griffin ; with the help
of David Witonsky and Edwin Willmore.
    p. cm. – (Research in mathematics education)
    Includes bibliographical references.
    ISBN-13: 978-1-59311-694-1 (pbk.) ISBN-13: 978-1-59311-695-8 (hardcover)
    1. Quadrilaterals. 2. Geometry, Plane. I. Usiskin, Zalman. II. Griffin, Jennifer.
III. Witonsky, David. IV. Willmore, Edwin. V. Title.
    QA482.U75 2007
    516'.154–dc22                      2007037395

Preparation of this monograph was partially supported by NSF Grant No. ESI-0333879
to the Center for the Study of Mathematics Curriculum, Barbara Reys, Director.
Opinions presented here are those of the authors and do not necessarily represent the
opinions of others at the center of the National Science Foundation.

Printed in the United States of America

# CONTENTS

# PREFACE

This monograph reports on an analysis of a small part of the mathematics curriculum, the definitions that are given to quadrilaterals. This kind of research, which we call *micro-curricular analysis,* is often undertaken by those who create curriculum, though normally not in as much detail as reported here, but it is not usually done systematically and it is rarely published. In fact, there are few avenues for the publication of such research. For this reason, few researchers in mathematics education undertake this kind of research.

One of the goals of the Center for the Study of Mathematics Curriculum, which has supported this study, is to encourage research into mathematics curriculum. The monograph series of the CSMC provides an avenue for the publication of results of research in the field.

## WHY QUADRILATERALS?

Many terms in mathematics education can be found to have different definitions in mathematics books. Among these are "natural number," "parallel lines," "congruent triangles," "trapezoid" and "isosceles trapezoid," the formal definitions of the trigonometric functions and absolute value, and implicit definitions of the arithmetic operations addition, subtraction, multiplication, and division.

Yet many teachers and students do not realize there is a choice of definitions for mathematical terms. And even those who realize there is a choice may not know who decides which definition of any mathematical term is better, and under what criteria. Finally, rarely are the mathematical implications of various choices discussed. As a result, many students misuse and otherwise do not understand the role of definition in mathematics (Edwards & Ward 2004).

In preparing this monograph, we were faced with a decision. Should we concentrate on the broader issues and use specific terms as examples for which there are choices of definition? Or should we devote our attention specifically to issues surrounding specific terms?

Reasonable arguments can be made for either approach. A monograph devoted to the broader issues enables a wider perspective on why these definitions differ, and would be very useful to those studying the growth of mathematical ideas in children as well as to those who are studying mathematics curriculum.

We have chosen in this monograph to take a middle ground and to examine a bit of mathematics for its definitions: the quadrilaterals. We do so because there is some disagreement in the definitions and, consequently, in the ways in which quadrilaterals are classified and relate to each other. The issues underlying these differences have engaged students, teachers, mathematics educators, and mathematicians. There have been several articles and a number of essays on the definitions and classification of quadrilaterals (see, for example, Craine & Rubenstein 1993; Pereira-Mendoza 1993; Prevost 1985; Keedy 1966; Maraldo 1980; the Math Forum Internet site, definitions and classification of quadrilaterals; and the Numericana internet site). But primarily we chose this specific area of definition in mathematics because it demonstrates how broad mathematical issues revolving around definitions become reflected in curricular materials. While we were undertaking this research, we found that the area of quadrilaterals supplied grist for broader and richer discussions than we had first anticipated.

## WRITING AND ACKNOWLEDGEMENTS

The original idea and the structure of this monograph are due to Zalman Usiskin. The task of examining textbook approaches to quadrilaterals and then organizing what was found was carried out first by David Witonsky in the fall of 2003 and early in 2004, and then by Jennifer Griffin through the rest of 2004. The final rounds of editing and formatting were completed by Edwin Willmore in the fall of 2005. The writing of the manuscript was done by Zalman Usiskin. He bears responsibility for the opinions presented here and also for any errors.

The authors wish to acknowledge six individuals who read a draft of the entire manuscript and provided exceedingly helpful suggestions for its improvement. In alphabetical order, they are: Thomas Banchoff, Brown University; Michael Battista, Michigan State University; James King, University of Washington; Peter Lappan, Michigan State University; Robert Reys, University of Missouri; and Doris Schattschneider, Moravian College (emerita).

The authors also wish to acknowledge the support of the management team of the CSMC: Barbara Reys, Doug Grouws and Ira Papick, University of Missouri; Glenda Lappan, Sandy Wilcox and Betty Phillips, Michigan State University; Chris Hirsch, Kate Kline, and Steve Ziebarth, Western Michigan University; and Iris Weiss, Horizon Research, Inc. Their continual encouragement throughout the writing was essential to the project from start to finish.

## AUDIENCES

The audiences for curriculum research are many and vary depending on the particular research questions being considered. For the classification of quadrilaterals, one audience consists of those who write any of the types of curriculum: curriculum goals and standards, textbooks, test specifications, and test items. These individuals must deal not only with the broad issues that underlie entire courses or full curricula but also with the micro-curricular issues of the wording of definitions, the selection of examples, and the construction and sequencing of exercises and problems.

A second audience is those who are engaged in preservice or inservice mathematics teacher education. The education of all mathematics teachers should include some attention to the issues surrounding the construction and selection of definitions.

Although some research in mathematics education can be of little interest to mathematics teachers, reactions from teachers to parts of this manuscript indicate that a large third audience is of teachers of geometry at the middle school, high school, and college levels. Teachers daily have to deal with the subtleties of the order and relationships among the topics they teach.

Other teachers of mathematics and those interested in mathematics or the learning of mathematics may also find the ideas in this monograph to be of use, even if they are not particularly interested in the topic of quadrilaterals. Our study has enlightened us with many surprises. We hope the same is true for you.

# INTRODUCTION

In 1984, three university students working for the University of Chicago School Mathematics Project (UCSMP) were sitting in a row at a long desk typing up the manuscript of the first UCSMP course, *Transition Mathematics*, when the master's degree student in the middle felt she found an error in the manuscript. The UCSMP manuscript from which she was working defined "trapezoid" as a quadrilateral with at least one pair of parallel sides. She felt a trapezoid could not have two pairs of parallel sides. She turned to her fellow workers on the right and the left, sure she was right. But she was disappointed. Both the student to her right and to her left said that, from what they knew, a trapezoid could have two pairs of parallel sides.

We found out then that the other two students had studied geometry from a book I had written with Art Coxford in the late 1960s, *Geometry—A Transformation Approach*. The student in the middle was typing a manuscript I had also written. Was I the only person who used this definition of trapezoid? The student in the middle, a mathematics major with a 4.0 GPA, was convinced that I was the only person who used this definition, and that my definition was wrong.

Most of people interested in this monograph are probably aware that there are two definitions of trapezoid that can be found in mathematics textbooks. The more common one in high school books is the one used in the geometry text that this student had studied from: (1) A trapezoid is a quadrilateral with *exactly* one pair of parallel sides. The less common one in high school books but a very common one in college texts is the one I have used in all my writing: (2) A trapezoid is a quadrilateral with *at least* one pair of parallel sides.

The difference between the word exactly and the word "at least" is quite significant. Since a parallelogram is universally defined as a quadrilateral with two pairs of parallel sides, the first definition means that *no* parallelograms are trapezoids. We call this an *exclusive* definition.

The second definition means that *all* parallelograms are trapezoids; it is *inclusive* in the same way that rectangles include squares. It makes properties of all trapezoids automatically apply to parallelograms. Also, with this inclusive definition, rectangles are special kinds of isosceles trapezoids.

The choice of definitions has implications for the sequencing of curriculum. If you use the inclusive definition, then you may want to have a formula

$$A = h\left(\frac{b_1 + b_2}{2}\right)$$

for the area of a trapezoid rather early, and develop the formula $A = bh$ for a parallelogram as a special case. If you use the exclusive definition, then you may never even connect the formula for a parallelogram with the formula for a trapezoid.

We use the inclusive definition of *trapezoid* in UCSMP middle and high school textbooks. We wrote about the definition in our teacher notes because we were aware that our use of this definition might bother some teachers. It also clearly bothered some students. Here is one letter we received about ten years ago, from a girl in Montana.

> I have found a most disturbing mistake in your math book, and let me tell you this is not the first time.
>
> On page 589, Lesson 13-5, Paragraph 2 (Trapezoids) in your *Transision* (sic) *Mathematics* you will find your mistake—A trapezoid is a quadrilateral that has at least one pair of parallel sides. I find this statement almost totally false a trapezoid has only one parallel side. Please write me with your opinion on this matter.

Around the same time, a teacher from Texas wrote us this thoughtful letter.

> "I want to inform you that there is a problem with the definition of trapezoid in the textbook *UCSMP Geometry* published in 1991 by Scott, Foresman and Company. [She goes on to state our definition and how under it, a parallelogram is a trapezoid.]
>
> According to the 1985 edition of *The World Book Encyclopedia*, Volume 16, page 2b, "A trapezoid is a quadrilateral with one set of parallel sides of unequal length." Three dictionaries and another geometry book give the same defini-

tion in different terminology. These five definitions clearly exclude parallelo-grams from the trapezoid classification.

In light of my research, your definition of trapezoid contradicts the standard one. Because of this, I disagree with your definition and hope it does not cause any confusion to students using this textbook. Students who have got-ten this far in mathematics without knowing the standard definition are apt to be seriously misled by yours. In future editions of this textbook, I would like to see this definition of trapezoid changed to comply with the commonly ac-cepted one. I would appreciate a reply from you regarding this problem.

It was a little surprising to get this kind of letter from a geometry teacher, because early in UCSMP *Geometry* we had a lesson titled "What is a 'cookie'?" precisely because we wanted both students and teachers to realize that defi-nitions are in theory arbitrary and that there is a choice in the definition of trapezoid, just as there is a choice in the definition of natural number (to some, 0 is a natural number; to others, it isn't), and in quite a few other mathematical terms.

But it was even more surprising to be sitting about seven years ago with a small group of mathematicians and mathematics educators while four of us were first planning a college mathematics textbook for high school teachers, and to hear a first-rate mathematician assert that there is only one correct definition of every mathematical term. And when pushed, this mathematician became even more insistent. For any mathematical term, only one definition is the right one.

The occasional letters we received from teachers and students about the definition of trapezoid made me curious about how many books used the inclusive definition and how many used the exclusive definition. So, about three years ago, we began looking at textbooks to see how they defined "trapezoid," and while we were at it, we decided to look at how textbooks defined all the quadrilaterals that they mentioned. We were interested in whether these definitions had changed over time, and we wound up ex-amining 101 high school geometry texts published in the United States from 1833 until the present. Some of these 101 books represent different editions by the same authors, and if all the definitions of the quadrilaterals were the same, we considered the two books as one. But if a definition of any type of quadrilateral changed, then we felt that the authors had recon-sidered all the definitions, so we treated it as a separate book. This gave us 86 different geometry texts. We also looked at 8 college-level geometry textbooks designed for mathematics majors and 16 college texts designed for a course in mathematics for elementary school teachers.

To many people, this would seem to be an unimportant bit of trivia, not worth more than a paragraph's attention. But, as the student and teacher letters indicate, and as the number of articles in the *Mathematics Teacher* and the number of reactions to a thread on a discussion group on the Math Forum confirm, this *particular* issue concerns a number of people quite strongly.

However, to deal with the particular issue (the definition of *trapezoid*) as if it were the only matter of interest would not have advanced the conversation. We attempt to show, by a detailed rendering of a group of definitions (the definitions of various types of *quadrilaterals*), that (a) mathematicians/ mathematics educators do *not* universally agree on definitions and (b) there are implications of these different definitions that go beyond purely mathematical considerations—that is, there are curricular implications of these definitions.

What we were unprepared for was the richness of the subject. The variety of equivalent definitions is surprising. *Trapezoid* is not the only term with non-equivalent definitions. *Kite* has an even more complex array of definitions. And both of these pale in comparison to the various definitions for *polygon*.

There is also a history of changing trends in definitions. There are connections of earlier definitions back to *Euclid*. And we could see the changes that occurred as a result of the 1923 report on secondary mathematics education. All this richness is in an area in which teachers often think they are teaching *the* agreed-on definitions of particular terms and no other definitions are as correct.

CHAPTER 1

# DEFINITIONS IN MATHEMATICS

In the world of language, a definition, such as one that we might find in a dictionary, is a statement of the meaning of a word, phrase or symbol. The same word may have many definitions with many nuances to each definition. These definitions are determined by how the word is used by others, not by how the authors or editors of the dictionary would like the word to be used. Consequently, these definitions may contradict each other.[1] Furthermore, a definition may include the word being defined,[2] thus stymieing the reader who has never previously seen the word.

Definitions in mathematics also encompass the fundamental idea of providing meaning to a word, phrase, or symbol. However, in writing mathematical discourse, the author does not have to select the same definition that others have used. Of course, if the author is solving a mathematical problem, then the author is obliged to use the terms in the same way that the problem-poser used them. Also, when an author is discussing the mathematical ideas of others, there is an obligation to use terms in the same ways that the others

---

1. For instance, when we "throw out" an idea, we may be putting it up for consideration, or we may be taking it out of consideration.
2. For instance, in the *Merriam Websters Tenth Collegiate Dictionary* (1994), the second listing of the word *orthodox* shows two definitions: "1. one that is orthodox 2. cap a member of an Eastern Orthodox church."

---

*The Classification of Quadrilaterals: A Study of Definition*, pages 1–8
Copyright © 2008 by Information Age Publishing

have used them. But if the author is creating mathematics, then the author has the liberty to choose or create the definition he or she prefers.

This freedom is contrary to the views that many people have about the language of mathematics. They think of the language of mathematics as being universal, and thus they view definitions in mathematics as being cast in stone. However, authors and others who have to think about every word that is written realize that there are options in definitions.

Whether or not an author is using an existing definition or creating a new one, there are additional constraints that distinguish definitions in mathematics from definitions in the world at large. In a single mathematical discourse a term that has a single definition and contradictory definitions cannot exist. A definition cannot include the word being defined. Furthermore, in theory a definition must include only terms previously defined or specifically designated as undefined. This theoretical criterion is quite impractical to follow in typical discourse, for to define articles such as "a" and "the," or logical terms such as "if" would pull the reader from the mathematical ideas being discussed, but the expectation is that key terms in the definition will have been dealt with previously.

For example, if "triangle" is defined in a geometrical discourse as "three-sided figure," the expectation would be that the words "side" and "figure" would have been explicitly defined previously so that their meanings are known. In this geometrical context, the word "three" would be viewed as a numerical idea whose meaning had been dealt with elsewhere and would not need to be explicitly defined.

The constraints on definitions in mathematics are of critical importance in the doing of mathematics. Definitions of terms form the basis from which properties of the terms are logically derived. Without careful unambiguous definitions, any mathematical results would be subject to question. For instance, without a careful definition of triangle, both a statement such as "the sum of the measures of the angles of any triangle is 180°" and any proof of the statement would be of little value. Mathematical definitions also provide the user with the means to name objects. Consider the objects drawn in Figure 1.1. Which of these are triangles? Without a careful definition of "triangle," the question cannot even be considered. In schoolbooks, only Figure 1.1b would be considered to depict a triangle. Although there are mathematical treatments of geometry in which each one of the figures 1a–1d might be a triangle,[3] in no treatment would any two of these be considered to be a triangle.

---

3. Figure 1.1a might depict a triangle on the surface of Earth, as studied in spherical geometry. As a plane figure, Figure 1.1a depicts a "Reauleaux Triangle," a figure made up of 120° arcs of circles with centers at the opposite endpoints. Figure 1.1c shows what is often called a "triangular region" and is often what we think of when we speak of the "area of a triangle." Figure 1.1d could depict a triangle in a finite geometry.

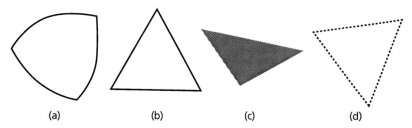

**Figure 1.1** Triangles.

Mathematics textbooks at the elementary school level often follow the standard dictionary conception of definition. From a mathematical perspective, the meanings of terms are often more accurately characterized as descriptions rather than as definitions.

At the high school level, and particularly in geometry textbooks, authors tend to provide mathematically precise definitions of terms. The definitions adhere reasonably closely to the principles mentioned above. However, many textbook definitions do not adhere to an additional criterion that is generally standard among mathematicians, that a definition not contain superfluous conditions. For instance, the most common definition of rectangle in U.S. geometry textbooks is "a parallelogram with four right angles." Yet a parallelogram with just one right angle can be shown to be a rectangle, making the criterion of the other three right angles superfluous to the definition.

Definitions of objects in mathematics are usually designed to classify them, first by identifying a category to which the object belongs, and then by indicating how this object is distinguished from other objects in that category. Sometimes authors insert redundant distinguishing characteristics to make it easier for students to deduce properties of the object.

The options open to authors are of three types:

(a) definitions that have different wordings but have the same defining conditions;
(b) definitions that have different defining conditions but are equivalent in the sense that they characterize the same objects;
(c) definitions that are different and not equivalent because they do not define exactly the same set of objects.

## AN INTRODUCTORY EXAMPLE—ISOSCELES TRIANGLES

To exhibit these types, and to show the kind of analysis we undertake in this monograph, we begin by considering a situation that is simpler than that of the quadrilaterals: the naming of certain triangles isosceles.

We view the following as the same but with different wordings, and thus examples of (a).

1. An *isosceles triangle* is a triangle with two congruent sides.
2. An *isosceles triangle* is a triangle with at least two congruent sides.
3. A triangle is *isosceles* if and only if it has two congruent sides.
4. A triangle is *isosceles* if and only if it has at least two congruent sides.

Some mathematics journals and books use the convention of substituting iff for if and only if when it is clear that a term is being defined. We view that as another example of (a).

The distinction between (a) and (b) is not always easy to determine. Compare the following definition of isosceles triangle with definition (1) above.

5. An *isosceles triangle* is a triangle with two sides of equal length.

Are (1) and (5) different but equivalent? Or are they merely different wordings of the same definition? The answer depends on the importance one wishes to give to the distinction between "congruent segments" and "segments of equal length." There is a choice.

It is usually easier to determine when we are dealing with option (c). Compare definition (5) with the following definition (6):

6. An *isosceles triangle* is a triangle with exactly two sides of equal length.

Under (6), which is the definition given in Euclid's *Elements*, an equilateral triangle is not isosceles. Under (5), the definition given in most of today's geometry texts, an equilateral triangle is isosceles. When two definitions like these exist, where one definition purposely excludes what the other definition includes, we call the one definition an *exclusive definition* and the other definition an *inclusive definition*.

A *scalene triangle* is consistently defined as a triangle in which no two sides are of the same length. Notice how different the Venn and tree diagrams are for scalene, isosceles, and equilateral triangles depending on which definition is chosen for isosceles triangle. (See Figures 1.2 and 1.3.)

The difference between definitions (5) and (6) has consequences beyond which triangles are isosceles. Consider the following two propositions:

**Proposition (1):**  Let $\overline{AB}$ be a segment and point M not on $\overline{AB}$ such that

$$AM = MB. \text{ Then } \triangle AMB \text{ is isosceles.}$$

**Proposition (2):**  Suppose $m\angle PQR = m\angle PRQ$. Then $\triangle PRQ$ is isosceles.

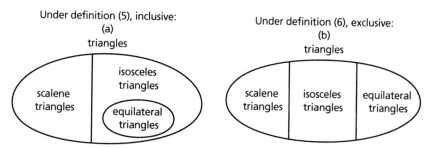

**Figure 1.2** Venn diagrams for scalene, isosceles, and equilateral triangles.

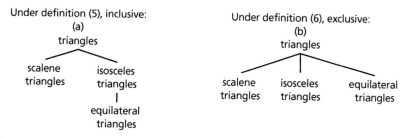

**Figure 1.3** Tree diagrams for scalene, isosceles, and equilateral triangles.

The truth of each proposition depends on the definition used for isosceles triangle. Under definition (5), both propositions are true. Under definition (6), both propositions are false because there is the possibility that the triangle might be equilateral.

We can modify the propositions so that they are true regardless of the definition of isosceles triangle. Here are three modifications of Proposition (1).

**Proposition (1′):** Let $\overline{AB}$ be a segment and point M not on $\overline{AB}$ be such that

AM = MB. Then ΔAMB is isosceles or equilateral.

**Proposition (1″):** Let $\overline{AB}$ be a segment and point M not on $\overline{AB}$ be such that

AM = MB. Then, if AB ≠ AM, ΔAMB is isosceles.

**Proposition (1‴):** Let $\overline{AB}$ be a segment and point M not on $\overline{AB}$ be such that

AM = MB. Then either AB = AM or ΔAMB is isosceles.

Each modification, (1′), (1″), and (1‴), contains an awkward conclusion that shows exceptional cases. Mathematicians today tend to prefer general theorems that do not have exceptions, when this is possible. This is why geometry books today uniformly use definition (5) for isosceles triangle even though Euclid's *Elements*, the most famous text of all time, used definition (6).

## HIERARCHIES, HIERARCHICAL CHAINS, AND PARTITIONS

An inclusive definition creates a link in a hierarchical chain from the more general to the more specific. With isosceles triangles, the inclusive definition (5) creates the chain

triangle $\supset$ isosceles triangle $\supset$ equilateral triangle.

Here the symbol $\supset$ means "includes." The chain could easily be made longer because every triangle is a polygon.

polygon $\supset$ triangle $\supset$ isosceles triangle $\supset$ equilateral triangle

It is possible to insert a link within an existing chain. A polygon is convex if, when two points on different sides of the polygon are connected by a line segment, all the other points of the segment are in the interior of the polygon. Thus some polygons are convex and others are not, but all triangles are convex polygons. This adds a link in the middle of the chain.

polygon $\supset$ convex polygon $\supset$ triangle $\supset$ isosceles triangle $\supset$
equilateral triangle

In contrast, an exclusive definition creates a partition of the more general object into a set of more specific objects. A partition of the set of all triangles is pictured in Figures 1.2b and 1.3b. Under definition (6), the set of all triangles is partitioned into scalene (no two sides of the same length), isosceles (exactly two sides of the same length), and equilateral triangles (all three sides of the same length). Another common partition of the set of all triangles is into acute, right, and obtuse triangles.

When partitions and hierarchical chains are combined into one diagram, the result is a hierarchy of the most general idea. Hierarchies of triangles are pictured in Figures 1.2b and 1.3b under definition (6). Figure 1.4 exhibits a hierarchy for polygons, using the inclusive hierarchy for isosceles triangles.

Figure 1.4 might give the impression that there is no classification of quadrilaterals. But, as we shall see, various hierarchies for quadrilaterals appear in the literature. The distinctions between inclusive and exclusive definitions and between hierarchical chains and partitions are useful for understanding these hierarchies.

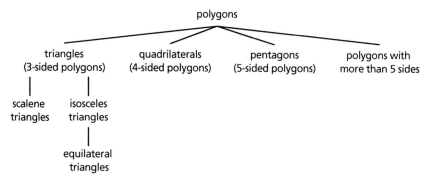

**Figure 1.4**   A hierarchy for polygons.

# CHAPTER 2

---

# QUADRILATERALS

## QUADRILATERALS AND QUADRANGLES

The natural extension etymologically from "triangle" as a three-sided poly-gon to one with four sides would be "quadrangle." This term is sometimes a synonym for *quadrilateral*.[1] "Quadrangle" sometimes refers to plane figures formed by connecting four points in any order as long as no three are col-linear (Figure 2.1a).[2]

Sometimes "quadrangle" refers to the figure consisting of all six segments that connect four points, where no three are collinear, also known as a com-plete quadrangle (Figure 2.1b). A complete quadrangle may also refer to the figure consisting of four points, no three of which are collinear, and the six lines they determine (Figure 2.2a). In contrast, a complete quadrilateral consists of four lines, no three of which are concurrent, and the six points they determine (Figure 2.2b).[3]

## QUADRILATERALS AS POLYGONS

Within mathematics, quadrilaterals provide the foundation for the study of areas of polygons and under curves, they help to describe how vectors

---

1. For example, see Coxeter, *Introduction to Geometry*, 23, 56, 57
2. See Schwartzman, *The Words of Mathematics*, 177.
3. See Gans, *Transformations and Geometries*, 252.

---

*The Classification of Quadrilaterals: A Study of Definition*, pages 9–20
Copyright © 2008 by Information Age Publishing

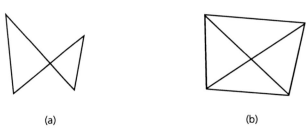

(a)                                             (b)

**Figure 2.1**   Some quadrangles.

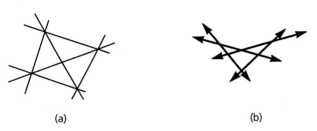

(a)                                             (b)

**Figure 2.2**   Complete quadrangle (a), and complete quadrilateral (b).

and complex numbers are added, and, of course, they are studied in geometry for their geometric properties. Furthermore, quadrilaterals in the
plane and the regions they determine have common applications outside
mathematics. They are the mathematical models of windows, roofs, copy
paper, faces of boxes, and a myriad of other objects. All quadrilaterals tes-
sellate, so they are basic figures in tiling and brick patterns. Starting with
the unit square, they are essential to understanding area, thus important
in determining sizes of land, surface areas and volumes of structures and
objects of all kinds. Because quadrilaterals are determined by the endpoints
of intersecting segments, they appear in kites, crossbows, frames, gates, and
other structures.

In school geometry texts, a *quadrilateral* is typically considered to be a
polygon with four sides even if that is not the wording found in the definition of "quadrilateral." This classification fits the etymology of "quadrilateral," from the Latin *quadri,* a combining form for "four," and *latus,* meaning "side." In Greek, such a figure was called a *tetragon,* literally meaning
"four angles." The actual definitions of "quadrilateral," however, show far
more variety. In the 86 geometry textbooks surveyed, we found six essentially different definitions.[4] On the next page we list those definitions as
Q(1) to Q(6) and give the number of texts having each definition.

---

4. One book had no definition and one book had two definitions.

| A quadrilateral is... | Number of texts |
|---|---|
| Q(1):  a four-sided polygon | 59 |
| Q(2):  the union of four line segments that join four coplanar points, no three of which are collinear, each segment intersecting exactly two others, one at each endpoint[a] | 10 |
| Q(3):  a closed, four-sided plane figure | 9 |
| Q(4):  a portion of a plane bounded by four sects | 4 |
| Q(5):  a closed figure with four straight line segments | 2 |
| Q(6):  a simple polygon with four sides[b] | 1 |

[a] A number of books that use definition Q(2) couch the definition in terms of named points, somewhat as follows: a figure made up of four coplanar points A, B, C, and D, no three of the points being collinear, and the segments $\overline{AB}$, $\overline{BC}$, $\overline{CD}$, and $\overline{DA}$ intersecting only at their endpoints.

[b] In this book, a "simple polygon" is a simple closed polygonal path, and a "polygonal path" is a path made up of several line segments.

Tracing these definitions back to the key words within them, we see that the meaning of "quadrilateral" depends variously on the meanings of one or more of the objects "figure," "polygon," "point," "plane," "endpoint," and "sect." Definition Q(2) also depends on the meanings of "coplanar" and "collinear," but this is true also of most of the books that use definition Q(1), for Q(2) incorporates one common definition of "polygon" into its definition. Appendix B shows that in 45 of the 46 books we examined, published since 1964, a quadrilateral is defined either directly or indirectly as a polygon with four sides.

The characterization of a quadrilateral as a polygon with 4 sides would seem to answer any question regarding what is and what is not a quadrilateral. However, the actual definitions given in high school geometry textbooks for "polygon" show even greater variety than those for "quadrilateral." In the 86 high school geometry textbooks that we examined from 1833 to 2004, we found 13 major categories of definitions, many with important variations that account for a total of 42 essentially different definitions! Furthermore, we found 7 more essentially different definitions in 16 mathematics texts for elementary school teachers that we examined and one more definition in the 8 college-level geometry texts we looked at, for a total of 50 essentially different definitions in 125 books.

Table 2.1 details 13 major categories of definition for "polygon" that we found in high school textbooks. This table describes each category, the number of textbooks (out of 86) in each category, the range of years in which definitions in this category appear, the variations within each category, and the number of textbooks that utilize each variation. Elaboration of each variation is found in Appendix A. The 13 major categories were

**TABLE 2.1  Classification of "Polygon" Definitions**

| Definition category | Definition approach or terminology | Number of books | Published range (years) | Variations | Number in variation |
|---|---|---|---|---|---|
| PG(1): | an orderable set of points along with the connecting line segments | 17 | 1961–1985 | A | 10 |
| | | | | B | 1 |
| | | | | C | 1 |
| | | | | D | 1 |
| | | | | E | 1 |
| | | | | F | 1 |
| | | | | G | 1 |
| | | | | H | 1 |
| PG(2): | a *plane figure bounded*... | 11 | 1893–1951 | A | 7 |
| | | | | B | 4 |
| PG(3): | a *closed plane figure*... | 11 | 1915–2004 | A | 2 |
| | | | | B | 1 |
| | | | | C | 1 |
| | | | | D | 1 |
| | | | | E | 1 |
| | | | | F | 1 |
| | | | | G | 1 |
| | | | | H | 1 |
| | | | | I | 1 |
| | | | | J | 1 |
| PG(4): | a *broken line*... | 9 | 1916–1963 | A | 8 |
| | | | | B | 1 |
| PG(5): | a *union* of segments or lines | 7 | 1961–2002 | A | 4 |
| | | | | B | 2 |
| | | | | C | 1 |
| PG(6): | a figure either *formed by*, *consists of*, or *made up of* segments; similar to PG(5) except the term *union* is not used. | 7 | 1978–1999 | A | 2 |
| | | | | B | 2 |
| | | | | C | 1 |
| | | | | D | 1 |
| | | | | E | 1 |
| PG(7): | a *portion of a plane*... | 6 | 1874–1948 | A | 5 |
| | | | | B | 1 |
| PG(8): | a *rectilinear figure*... | 4 | 1833–1901 | A | 2 |
| | | | | B | 1 |
| | | | | C | 1 |
| PG(9): | a *simple*... | 3 | 1984–1997 | A | 1 |
| | | | | B | 1 |
| | | | | C | 1 |
| PG(10): | a *closed figure, bounded*... | 2 | 1998–2003 | — | 2 |
| PG(11): | a *plane figure*... | 1 | 1991 | — | 1 |
| PG(12): | a *closed two dimensional shape* formed by three or more line segments | 1 | 1997 | — | 1 |
| PG(13): | a *closed figure with straight lines as sides* | 1 | 1958 | — | 1 |
| no definition | | 6 | | | |

identified based on key words and or the broad approach used in the definitions. Within a category there is still variation with respect to some of the conditions and nomenclature used. For example, consider the following two definitions, both grouped into category $P(1)$,

> **PG(1)A:** let $P_1, P_2, \ldots, P_n$ be $n$ distinct points in a plane, $n \geq 3$. The union of segments $\overline{P_1 P_2}, \ldots, \overline{P_n P_1}$ is a polygon if
>
> 1. No two of the segments intersect except at their endpoints.
> 2. No two segments with a common endpoint are collinear.

and,

> **PG(1)F:** let $P_1, P_2, \ldots, P_n$ be $n$ distinct points in a plane, $n > 2$. The union of segments $\overline{P_1 P_2}, \ldots, \overline{P_n P_1}$ is a polygon if no three consecutive points are collinear.

Upon initial inspection, PG(1)A and PG(1)F are very similar. They both construct a set of distinct ordered points in a plane along with the connecting line segments. We considered $n \geq 3$ and $n > 2$ as a trivial difference and insignificant. However, condition 2 of PG(1)A, "No two segments with a common endpoint are collinear," and the sole condition of PG(1)F, "no three consecutive points are collinear," differ in that the former is a condition on segments while the latter is a condition on the points involved. Even though these are equivalent conditions in this context, the difference is enough for us to consider these definitions as variations within a category. Furthermore, condition 1 of PG(1)A excludes self-intersection, while PG(1)F has no equivalent condition. The resulting sets of objects defined as polygons are not identical, so the definitions are not equivalent.

However, this difference did not preclude their both being classified under PG(1). When these two definitions are contrasted with a definition from another category,

> **PG(2)B:** plane closed figure whose boundary consists of straight lines only.

the similarities between PG(1)A and PG(1)F are evident.

Graph 2.1 reveals that definition categories of polygon go in and out of favor. For instance, PG(2), PG(7), and PG(8) are the only definition categories appearing from 1833 to 1915, and all three stop appearing by 1960. Only one of the eight definition categories that appears after 1970 appears before 1960, PG(3). And most interesting of all, the most populous definition cat-

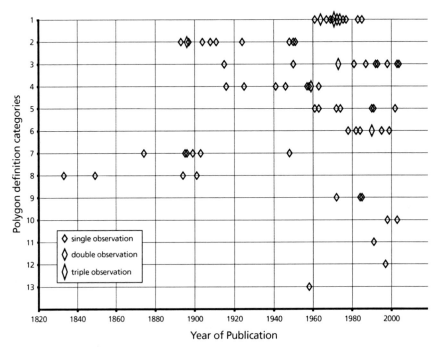

**Graph 2.1** Polygon definition categories over time.

egory PG(1), was used in 17 of 33 books we surveyed that were published in the period 1961 to 1985, but does not appear before or after that period.

In order to understand these 13 definition categories, we compared the essentially different definitions by considering what types of drawings they allow as polygons. Figure 2.3 shows eight objects which are possible polygons, seven of which are contentious among the definitions in high school texts, along with the number of published definitions that allow each object as a polygon.[5]

Of the seven types of figures in Figure 2.3 that are allowed as polygons by some definitions 2.3(b), 2.3(d), and 2.3(e) do not apply to quadrilaterals or other polygons with less than six sides. Still, it is important to note that type 2.3(d) qualifies as a polygon in 50 of the 86 textbooks surveyed. In these books, a collection of two disjoint triangles qualifies as a hexagon, and three squares qualifies as a dodecagon. It would seem that such inclusions had occurred to neither authors, teachers, nor students.

---

5. Six books had no definition, leaving 80 as the maximum number of books considering each object a polygon..

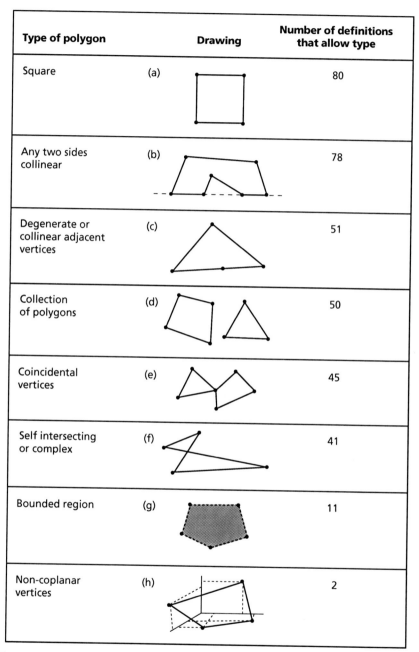

| Type of polygon | Drawing | Number of definitions that allow type |
|---|---|---|
| Square | (a) | 80 |
| Any two sides collinear | (b) | 78 |
| Degenerate or collinear adjacent vertices | (c) | 51 |
| Collection of polygons | (d) | 50 |
| Coincidental vertices | (e) | 45 |
| Self intersecting or complex | (f) | 41 |
| Bounded region | (g) | 11 |
| Non-coplanar vertices | (h) | 2 |

**Figure 2.3**  Types of polygons allowed in definitions.

The word "closed," found in 33 textbook definitions spanning 8 definition categories, is rarely defined, and in most cases it is taken as understood. When it is either discussed or defined, the meaning varies dramatically. Four examples of what is meant by closed are:

i) *a closed line encloses a portion of a plane,* (Seymour and Smith 1941)

ii) *closed is the idea that if we start at any point on the figure and 'move' along the sides, we will at some time arrive back at the starting point,* (Keenan and Dressler 1985)

iii) *a simple closed curve is a plane curve which can be drawn without lifting the pencil, and returning to the starting point without tracing any point more than once except the starting point,* (Bumby and Klutch 1985)

iv) *a closed figure is one that can be traversed by starting at any point of it, and moving continuously along the lines of the figure in order, returning to the same point without passing twice over any portion of the figure.* (Holgate 1901)

Sometimes "closed" allows self-crossing paths as in the first two, sometimes it disallows them as in the last two. Objects like those in Figure 2.3(d) would be closed according to the first two but not the last two. (After all, the first description of closed says nothing about whether the portion of the plane must be connected.) Objects like Figure 2.3(c) would be closed according to the first three definitions, but not the last one. (Regarding the third definition, if we choose the coincidental vertex of Figure 2.3(e) as our starting point, the figure can be drawn with only the starting point being traced more than once.) As a result of such varied meanings for closed, even identically phrased definitions can include different objects as polygons.

The reader whose experience includes no definition other than PG(1)A may be bothered both by the inconsistency shown among these definitions and by the inclusion of figures that in many books are specifically asserted *not* to be polygons. Yet, as can be seen by an examination of Appendices A and B, the existence of a variety of definitions for types of quadrilaterals, including non-equivalent definitions is the rule rather than the exception.

Type 2.3(c), in which a 4-sided polygon is also a triangle, is a polygon in over half of the books we examined. Such a figure is often called a *degenerate quadrilateral.* With a dynamic geometry drawing program such as *Geomenter's Sketchpad* or *Cabri*, moving a vertex of any polygon with *n* sides (an *n*-gon) can move consecutive sides to be collinear segments and thus create a polygon with one less side (an (*n*-1)-gon). Study of these figures can be enlightening, because some theorems for the *n*-gon remain valid for the (*n*-1)-gon. In Chapter 7 of this monograph we discuss degenerate quadrilaterals.

Allowing polygons to include non-coplanar figures as in type 2.3(h) also gives rise to extentions of theorems, that may have been thought to apply only to plane figures. For instance Varignon's Theorem, that the figure

**TABLE 2.2  Equivalent Definitions by the Objects they Allow to be Polygons**

| Objects allowed from Figure 2.3 | | | | | | | | Polygon definitions | Number of books | Years of publication |
|---|---|---|---|---|---|---|---|---|---|---|
| a | b | c | d | e | f | g |   | 2A, 2B, 7A, 8A | 18 | 1833–1951 |
| a | b |   |   |   |   |   |   | 1A, 1B | 11 | 1964–1985 |
| a | b | c | d | e | f |   |   | 3E, 3F, 3G, 3J, 5A, 12, 13 | 10 | 1915–2002 |
| a | b |   | d |   |   |   |   | 3D, 5B, 6A, 6B, 6D | 8 | 1973–1999 |
| a | b | c | d |   |   |   |   | 3A, 6E, 10, 11 | 6 | 1972–2003 |
| a | b | c |   | e | f |   |   | 1C, 4A | 6 | 1916–1961 |
| a | b | c |   |   |   |   |   | 1H, 8B, 9A, 9B, 9C | 5 | 1901–1985 |
| a | b |   | d | e |   |   |   | 3B, 3H, 3I | 3 | 1981–2004 |
| a | b |   |   | e | f |   |   | 4A, 4B | 3 | 1957–1959 |
| a | b | c | d | e |   |   |   | 3C, 8C | 2 | 1894–1993 |
| a | b | c | d |   | f |   |   | 6C | 1 | 1978–1990 |
| a | b | c |   |   |   |   | h | 1E | 1 | 1983 |
| a |   |   | d |   |   |   |   | 5C | 1 | 1972 |
| a |   |   |   |   |   |   |   | 1D | 1 | 1965–1972 |
| a | b |   |   |   | f |   |   | 1F | 1 | 1967 |
| a | b | c |   | e |   |   | h | 1G | 1 | 1964 |
| a | b |   | d |   | f |   |   | 4A | 1 | 1963 |
| a | b | c |   | e | f | g |   | 7B | 1 | 1948 |

formed by connecting midpoints of the sides of a quadrilateral in order is a parallelogram, is true whether or not all sides of the quadrilateral are coplanar.

Table 2.2 shows that there are 18 classes of equivalent definitions, as determined by what figures they allow to be polygons. The most common class of equivalent definitions found in our study is also the most inclusive, since it allows seven of the eight objects in Figure 2.3. However, the four definitions that compose this class do not appear after 1951. Table 2.2 also shows that 4 non-equivalent definitions of polygon are found in books published in the last 10 years. All of these allow Figure 2.3(d).

## TYPES OF QUADRILATERALS

It is natural to distinguish various types of quadrilaterals by their properties. Six types of quadrilaterals are found in virtually all school geometry texts: *parallelograms, rectangles, rhombi* (*rhombuses* or *rhombs*), *squares, trapezoids,* and *isosceles trapezoids.* Examination of high school geometry texts and texts for elementary school teachers indicates that a seventh type, the *kite*, is becoming more common.

Appendices A and B display the range of definitions for these seven types of quadrilaterals as found in 86 high school geometry texts published from 1833 to 2004. In the texts published since 1930, we find essentially four different hierarchies of quadrilaterals, with the choice of hierarchy depending on the selection of inclusive or exclusive definitions for *trapezoid* and *kite*. Specifically, the hierarchies differ because of different answers given to the following questions:

1. Are trapezoids defined so that every parallelogram is a trapezoid?
2. Are isosceles trapezoids defined so that every rectangle is an isosceles trapezoid?
3. Are kites defined so that every rhombus is a kite?
4. Are kites defined so that a kite can be nonconvex?

In Figure 2.4, we show the choices of hierarchies determined by the answers to the first three questions. In this figure a *thick* line means that the figure below is *always* considered to be a special case of the figure above, while a *thin* line means that the figure below is *sometimes* considered to be a special case of the figure above. The hierarchy is the same whether or not a kite must be convex; all that changes is which figures can be kites.

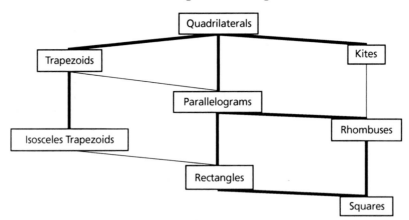

**Figure 2.4** Hierarchies of quadrilaterals.

## QUADRILATERALS IN EUCLID

The hierarchy of quadrilaterals in Figure 2.4 contrasts markedly with the classification of quadrilaterals found in Book I of Euclid's *Elements*. At the beginning of Book I a large number of terms are defined, including five types of quadrilaterals:

> Definition 22. Of quadrilateral figures, a square is that which is both equilateral and right-angled; an *oblong* that which is right-angled but not equilateral; a *rhombus* (from Greek: ρομβοσ) that which is equilateral but not right-angled; and a *rhomboid* (from Greek: ρομβοειδεσ) that which has its opposite sides and angles equal to one another but is neither equilateral nor right-angled. And let quadrilaterals other than these be called *trapezia.*[6]

The resulting hierarchy is one in which none of the five named types of quadrilaterals is a special kind of any other type (Figure 2.5).

This partition of the set of quadrilaterals is as exclusive a set of definitions of quadrilaterals as possible. It is significantly different from the classification found in today's texts.

Euclid never mentions oblongs, rhombuses, or rhomboids, and later in the *Elements* he does introduce parallelograms (without definition!) in Proposition 34 of Book I: "In parallelogramic areas, the opposite sides and angles are equal to one another and the diagonal bisects the angles."[7] Many of the remaining propositions of Book I are about parallelograms. It seems as if the meaning of parallelogram was so clear from its Greek roots (see Chapter 3) that no definition was necessary. Despite the definitions given at the start of Book I, Euclid treats rhombuses and rhomboids as special kinds of parallelograms, and he introduces rectangular parallelograms that include both oblongs and squares.

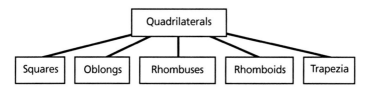

**Figure 2.5** Euclid's hierarchy of quadrilaterals.

6. Euclid. *Elements,* Volume 1: 154.
7. Ibid., 323.

# CHAPTER 3

# PARALLELOGRAMS

From the standpoint of definitions of the individual types of quadrilaterals, the parallelogram is the simplest to analyze. All but one of the high school geometry texts published since 1930 that we examined define parallelogram using the definition P(1) below.[1] This definition is clearly the same as the one used but not explicitly stated in Euclid's *Elements*.

| A parallelogram is... | | Number of texts[a] |
|---|---|---|
| P(1): | a quadrilateral with two pairs of parallel sides. | 84 |
| P(2): | a trapezoid with two pairs of parallel sides. | 1 |

[a]  One text gave no definition.

For students, a picture may be at least as powerful as a definition. In this regard, it would be interesting to study the drawings of parallelograms that are found in texts. Though we have not done an in-depth study, it is our impression that most books give one drawing of a parallelogram by its definition, and that this parallelogram has one pair of horizontal sides (Figure 3.1a). We believe it is rare to find a book that shows a parallelogram whose sides differ greatly in length and are positioned to be neither horizontal nor vertical (Figure 3.1b). It is also rare to find a book that emphasizes the inclusive nature of parallelograms—that is, that all rhombuses, squares,

---

1.  See Appendices A–C.

---

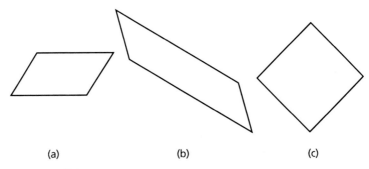

|  (a)  |  (b)  |  (c)  |

**Figure 3.1**  Parallelograms.

and rectangles are parallelograms—by showing one of these special types of parallelograms (such as in Figure 3.1c) along with the definition of parallelogram and before these types themselves have been defined.

As can be seen in Appendices A and B, parallelograms are unique among the various types in having one definition so dominant in textbooks. It may be that the term "parallelogram" so obviously includes "parallel" that any definition not involving parallelism is uncomfortable. The word parallelogram is from the Greek "parallelos" *parallel* and "gram" *something written,* especially a letter or other symbol or a line. Many other words in and out of mathematics use the same root as "gram," including graph, the metric unit gram (from "gramma," a small weight such as that used to create early letters in clay), tangram, diagram, program, and grammar.

Although only one definition dominates books, other definitions could be given for *parallelogram* that would be equivalent to it. From the standpoint of logic, the requirement for a definition to be equivalent to the one given above is that the defining condition yields the same figures, that is, that the defining condition describes quadrilaterals with two pairs of parallel sides, and only such figures. Here are four of the many possible equivalent definitions.[2]

A quadrilateral is a *parallelogram* (Figure 3.2) if and only if

(a)  both pairs of opposite sides have the same length.
(b)  both pairs of opposite angles have the same measure.
(c)  its diagonals have the same midpoint (bisect each other).
(d)  it possesses rotation symmetry.

---

2.  Some other possible defining conditions are (e) one pair of sides is parallel and of equal length; (f) two pairs of adjacent angles are supplementary; and (g) a diagonal divides the quadrilateral into two congruent triangles with the same orientation.

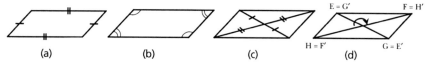

**Figure 3.2** Some possible defining conditions for parallelograms.

## Showing That Definitions Are Equivalent

Although it might be strange to use any of the definitions (a) through (d) as the definition of *parallelogram* (in place of the customary definition as a quadrilateral with two pairs of parallel sides), the defining conditions in definitions (a) to (b) can be used to describe hierarchies of quadrilaterals based on properties of (a) sides, (b), angles, (c) diagonals, and (d) symmetry. For this reason, we take time here to show that these definitions are equivalent.

To show that two definitions are equivalent, we must show that each defining condition implies the other. Here is a proof that the defining conditions (a) and (b) of *parallelogram* stated above are equivalent. Our strategy is to show (a) ⇒ (b) ⇒ (P1) ⇒ (a).

**(a) ⇒ (b):** We have to show that if, in a quadrilateral, both pairs of opposite sides have the same length, then both pairs of opposite angles have the same measure. Suppose that the quadrilateral is PQRS and that PQ = RS and PS = RQ, as in Figure 3.3. Draw $\overline{QS}$.

Then, since QS = QS, ΔPQS ≅ ΔRSQ by the SSS triangle congruence proposition. Now the corresponding angles P and R of these triangles have the same measure. But P and R are also opposite angles of the quadrilateral. The same argument can be repeated drawing the diagonal $\overline{PR}$ instead of $\overline{QS}$. Then the corresponding angles ∠Q and ∠S of the congruent triangles PQR and RSP have the same measure.

**(b) ⇒ (a):** We have to show that if, in a quadrilateral, both pairs of opposite angles have the same measure, then both pairs of opposite sides have the same measure. Suppose again that the quadrilateral is PQRS and that m∠P = m∠R and m∠Q = m∠S. We will make use of the fact that

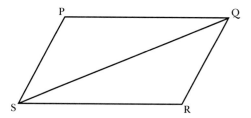

**Figure 3.3** Quadrilateral PQRS with diagonal.

the sum of the angles of a triangle is 180°, and so the sum of the mea-sures of the angles of a quadrilateral is 360°. Since $m\angle P = m\angle R$ and $m\angle Q = m\angle S$ and $m\angle P + m\angle R + m\angle Q + m\angle S = 360°$, then by substitution $m\angle P + m\angle P + m\angle Q + m\angle Q = 360°$. Thus $2(m\angle P + m\angle Q) = 360°$, from which $m\angle P + m\angle Q = 180°$. Since P and Q are interior angles on the same side of the transversal $\overline{PQ}$, and they are supplementary, the sides $\overline{PS} \mathbin{//} \overline{QR}$. Thus the alternate interior angles PSQ and RQS have the same measure. A similar argument with angles $\angle P$ and $\angle S$ shows that $\overline{PQ} \mathbin{//} \overline{RS}$ and that alternate interior angles $\angle PQS$ and $\angle RSQ$ have the same measure. This shows (b) $\Rightarrow$ P(1). Since QS = QS, by the ASA triangle congruence proposi-tion, $\triangle PQS \cong \triangle RSQ$. So the corresponding sides PQ = RS and PR = QS. So ((b) and P(1)) $\Rightarrow$ (a).[3]

Since (a) $\Rightarrow$ (b), (b) $\Rightarrow$ P(1), and P(1) $\Rightarrow$ (a),
(a) is also equivalent to P(1).

The defining conditions in definitions (a) through (c) are properties of parallelograms that are mentioned in most high school geometry texts and often proved. The property in (d) is in a minority of books so we describe it here. In general, a figure F possesses rotation symmetry if and only if there is a rotation R under which R(F) = F. That is, the rotation must map the fig-ure onto itself. With quadrilaterals, the only rotations that can possibly map a figure onto itself have their center at the intersection of the diagonals and have a magnitude of either 90° (a *quarter-turn*) or 180° (a *half-turn*). Such quadrilaterals are depicted in Figure 3.4, where in each case point O is the center of the rotation.

For a quarter-turn R to map the quadrilateral ABCD onto itself (Fig-ure 3.4a), each vertex in turn must be the image of the preceding one. So if R(A) = B, then we must have R(B) = C, R(C) = D, and R(D) = A. Or, if R(A) = D, then R(D) = C, R(C) = B, and R(B) = A. We say that ABCD has *four-fold rotation symmetry*, or sometimes just *four-fold symmetry*.

When a quadrilateral has four-fold symmetry, its diagonals are perpen-dicular and have the same length, and it is rather easy to show that the figure must be a square.

For a half-turn R to map the quadrilateral EFGH onto itself (Figure 3.4b), opposite vertices must be the images of each other under R. So we

---

3.  The proof that (b) $\Rightarrow$ (a) given here relies on the sum of the measures of a triangle being 180°. This means that the proof implicitly uses the parallel postulate of Euclidean geometry. It is possible to prove (b) $\Rightarrow$ (a) without any dependence on the parallel postulate. The proposition that both pairs of op-posite sides of a quadrilateral are equal in length if and only if both pairs of opposite angles of the quadrilateral are equal in measure is true in hyperbolic non-Euclidean geometry as well as in Euclidean geometry.

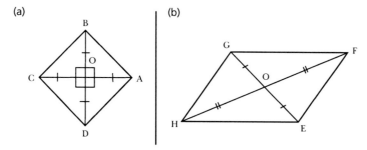

**Figure 3.4**  Quadrilaterals that possess rotation symmetry.

must have $R(E) = G$, $R(F) = H$, $R(G) = E$, and $R(H) = F$. We say that EFGH possesses *two-fold rotation symmetry* or *two-fold symmetry* or that EFGH is *point-symmetric* about O. This means that $OE = OG$ and $OF = OH$, i.e., that the diagonals bisect each other.[4] So the defining conditions (c) and (d) of a parallelogram are equivalent to each other: if the diagonals of a quadrilateral bisect each other, then the quadrilateral has rotation symmetry, and conversely.

We have so far shown that, as defining conditions for a special type of quadrilateral,

$$(a) \Leftrightarrow (b) \Leftrightarrow P(1)$$

and that
$$(c) \Leftrightarrow (d).$$

To show that all of these are equivalent, we only need to show that one of the conditions of the first group is equivalent to one of the conditions of the second group. We show here that $(a) \Leftrightarrow (c)$.

**(a)** $\Rightarrow$ **(c):** Suppose that opposite sides of quadrilateral WXYZ have the same length, that is, that $WX = YZ$ and $WZ = XY$ (Figure 3.5). Then $\triangle WXY \cong \triangle YZW$ by SSS congruence, so $m\angle WYX = m\angle YWZ$. Also, $\triangle XYZ \cong \triangle ZWX$ by SSS congruence, so $m\angle YXZ = m\angle XZW$. That is, each diagonal splits the quadrilateral into congruent triangles and corresponding angles in those triangles have the same measure. Now consider triangles YCX and WCZ formed by the diagonals. These triangles are congruent by ASA congruence and so $CX = CZ$ and $CY = CW$. Thus the diagonals bisect each other. This shows that $(a) \Rightarrow (c)$.

**(c)** $\Rightarrow$ **(a):** Suppose the diagonals of WXYZ bisect each other. Then $CX = CZ$ and $CY = CW$. Then, because the vertical angles $\angle WCZ$ and $\angle YCX$ have the same measure, $\triangle CXY \cong \triangle CZW$ by SAS congruence. So the corresponding sides $\overline{WZ}$ and $\overline{XY}$ have the same length. Also, $\triangle CXW \cong \triangle CZY$ by

---

4. Notice that a figure with 4-fold symmetry must possess 2-fold symmetry, because the composite of two 90° rotations with the same center is a 180° rotation with that center.

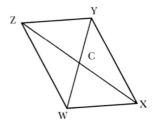

**Figure 3.5**   Quadrilateral WXYZ with diagonals intersecting at C.

SAS congruence, so WX = XZ. So the opposite sides of the figure have the same length. Thus (c) ⟺ (a) and (a) ⟺ (b) ⟺ P(1) ⟺ (c) ⟺ (d). That is, all of the defing conditions (a), (b), (c), and (d) are equivalent to P(1); they yield the identical figures.

## The Hierarchy Underneath Parallelograms

Just as there is no disagreement regarding the definition of parallelogram, and which quadrilaterals should be called parallelograms, there is no disagreement among today's textbook authors regarding which special types of quadrilaterals are always parallelograms. Rectangles, rhombuses, and squares are universally viewed as parallelograms, as depicted by the thick lines in the hierarchy diagram of Figure 2.4, partially repeated here as Figure 3.6.

Euclid's rhomboids, figures that we would describe today as "parallelograms that are neither rectangles nor squares," were defined in some U.S. textbooks prior to 1930 but none since. Likewise, the drawing of a parallelogram that is shown to students as "a general parallelogram" is most likely to be that of a rhomboid. As a result, students may not realize that rectangles,

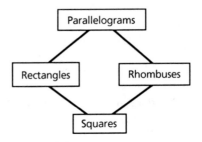

**Figure 3.6**   Hierarchies of parallelograms.

rhombuses, and squares are parallelograms.

# CHAPTER 4

# TRAPEZOIDS

Although the same set of figures are classified as parallelograms in all books, the place of parallelograms in the quadrilateral hierarchy depends on the definition given to trapezoids. In the 86 high school geometry texts we examined, we found two different[1] defining conditions for trapezoid. We call these T(I) for the inclusive definition and T(E) for the exclusive definition (See Figure 4.1). The numbers following the definitions are the numbers of books using each definition.[2]

| A trapezoid is... | Number of texts |
|---|---|
| T(E): a quadrilateral with exactly one pair of parallel sides. | 76 |
| T(I): a quadrilateral with at least one pair of parallel sides. | 8 |

After pointing out the two definitions of trapezoid mentioned above, Schwartzman[3] observes:

---

1. By "different," we mean "with a different defining condition." The wording of the definitions may differ substantially. For instance, Herberg and Orleans 1958 define a trapezoid as "a 4-gon formed by two parallels and two transversals."
2. Two books did not give a definition.
3. Schwartzman, op. cit., p. 225.

---

*The Classification of Quadrilaterals: A Study of Definition*, pages 27–32
**27**

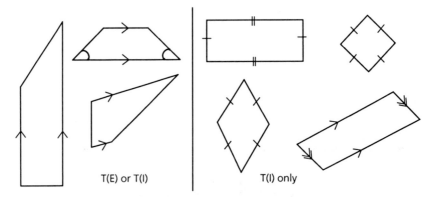

T(E) or T(I)    T(I) only

**Figure 4.1** Trapezoids according to definitions T(I) or T(E).

> The situation is further confused by the fact that in Europe a trapezoid is defined as a quadrilateral with no sides equal. Even more confusing is the existence of the similar word *trapezium,* which in American usage means "a quadrilateral with no sides equal," but which in European usage is a synonym of what Americans call a trapezoid. Apparently to cut down on the confusion, *trapezium* is not used in American textbooks.

We found *trapezium* defined only in those U.S. textbooks published before 1930 (see Appendix B).

The word *trapezoid* comes from the Greek word "trapeza" meaning *table* and the suffix "-oid" meaning *resembling.* Perhaps this comes from the realization that when a rectangular table is viewed from the front (that is, in one-point linear perspective from a front view), it has the shape of a quadrilateral with its front and back sides parallel but its other sides not parallel (see Figure 4.2). Other words with one of these roots are trapeze, cardioid (a heart-shaped curve), asteroid (like a star), humanoid, etc.

Parallel sides of a trapezoid are called *bases* and the distance between bases is the height of the trapezoid for those bases. So a trapezoid has exactly one pair of bases and a unique altitude under the exclusive definition but may have two pairs of bases and a height for each pair under the inclusive definition.

**Figure 4.2** A visualization of a rectangular table rotated to project as a trapezoid.

Under the exclusive definition, parallelograms are not trapezoids. Under the inclusive definition, they are. Thus the decision one makes in choosing a definition for trapezoid is precisely whether one wishes to include parallelograms in the trapezoid family. Figure 4.1 displays this fact by showing rectangles, rhombuses, squares, and parallelograms as trapezoids at the right, under T(I). The quadrilaterals at the left in Figure 4.1 are trapezoids under either definition T(E) or T(I).

Because trapezoids are high up in the hierarchy of quadrilaterals, and because parallelograms are pivotal in the discussion of properties of quadrilaterals, the choice of definition of trapezoid has implications for the derivation of properties not only of trapezoids but also of parallelograms, isosceles trapezoids, rectangles, rhombuses, and squares.

## CHOOSING BETWEEN EXCLUSIVE AND INCLUSIVE DEFINITIONS

We have noted that to identify a general parallelogram, one usually draws a parallelogram that is distinguished from any special types and any proof has to be general enough to include all such parallelograms. Likewise, to prove any property of all trapezoids, under either the inclusive or exclusive definition, the proof has to be broad enough to include all trapezoids, and the figure that is used as an aid in the proof would normally be chosen from those at the left in Figure 4.1, not those at the right. In this sense, the exclusive definition of trapezoid serves to distinguish those quadrilaterals that have one pair of parallel sides and are not parallelograms.

One way to decide whether the inclusive or exclusive definition is more appropriate is to ask whether a proof of a property of trapezoids using an exclusive definition has to be modified to apply to the figures in an inclusive definition. If no modification is necessary, then it is an unnecessary waste of energy to restate the proof for the other figures and the exclusive definition is inefficient. But if a modification is necessary, then it provides more weight in the favor of the exclusive definition.

For example, the most common theorem about trapezoids in geometry textbooks is a formula for calculating their areas:

**Theorem (T-I1):** In a trapezoid with bases of lengths $b_1$ and $b_2$, height $h$, and area $A$,

$$A = \frac{1}{2}h(b_1 + b_2).$$

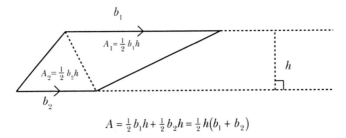

$$A = \tfrac{1}{2}b_1 h + \tfrac{1}{2}b_2 h = \tfrac{1}{2}h\left(b_1 + b_2\right)$$

**Figure 4.3**   Derivation of the area formula $A = \tfrac{1}{2}h(b_1 + b_2)$ for a trapezoid.

A common proof of this theorem is to separate the trapezoid into triangles and add their areas, and then use the distributive property of algebra, as shown in Figure 4.3. This proof works, without any changes, for parallelograms whether or not they are considered to be trapezoids. It suggests that the inclusive definition might be the more natural definition.

Other properties of trapezoids defined exclusively (we call these trapezoids-E) that also apply to parallelograms (we call these trapezoids-I) are the following (Figure 4.4):

**Theorem (T-I2):**   In a trapezoid-I, there are two pairs of adjacent supplementary angles.

**Theorem (T-I3):**   In a trapezoid-I, the diagonals form a pair of similar non-adjacent triangles.

**Theorem (T-I4):**   In a trapezoid-I, the point of intersection of the diagonals splits the diagonals proportionally.

On the other hand, there are theorems that apply only to trapezoids that are not parallelograms. Consider the following.

**Theorem (T-E1):**   In a trapezoid-E, the segment connecting the midpoints of the diagonals is parallel to two of the sides of the trapezoid and equal in length to half the difference in lengths of the two bases.

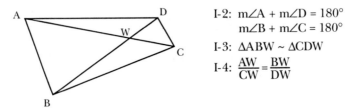

I-2:  $m\angle A + m\angle D = 180°$
      $m\angle B + m\angle C = 180°$

I-3:  $\triangle ABW \sim \triangle CDW$

I-4:  $\dfrac{AW}{CW} = \dfrac{BW}{DW}$

**Figure 4.4**   Properties of trapezoids defined inclusively.

In a parallelogram there is no segment connecting the midpoints of the diagonals, because in a parallelogram the midpoints of the diagonals coincide. So the theorem, as stated, has no meaning for parallelograms. However, we could consider the "segment" joining the midpoints to be just a single point. Although this is not a segment in the usual sense, if we think of this single point as a segment whose endpoints coincide, then its length is 0. Since the bases in a parallelogram are congruent, the difference of their lengths is 0, so half the difference in their lengths is also 0, indicating that the theorem interpreted in this way is true for parallelograms.

This type of special case, where a figure with some measure collapses into an object with zero measure, is not unusual in mathematics. For instance, a zero vector has coinciding initial point and endpoint, and its length is considered to be zero. A circle is sometimes considered as an ellipse whose two foci coincide (at the center of the circle).

There is a fundamental property of trapezoids-E that is not true for all trapezoids-I. All trapezoids-E can be constructed through a truncation of some triangle by a line parallel to one of the triangle's sides (Figure 4.5). This property comes straight from the exclusive definition: the lines containing the sides of the trapezoids that are not bases are intersecting lines. The trapezoids formed fall under both definitions, however all trapezoids-E can be constructed by this method of triangle truncation, whereas exactly those trapezoids-I that have two sets of parallel sides (parallelograms) cannot be constructed through this method.

> **Theorem (T-E2):** Let ABCD be a trapezoid-E whose bases are $\overline{AB}$ and $\overline{CD}$. Let P be the intersection of the lines containing sides $\overline{AD}$ and $\overline{BC}$. Then $\triangle PAB \sim \triangle PDC$ and so
> $$\frac{PD}{PA} = \frac{PC}{PB}.$$

In a treatment where trapezoids are defined inclusively, Theorem E-2 would have to be reworded to begin with a restriction as in "Let ABCD be a trapezoid that is not a parallelogram..."

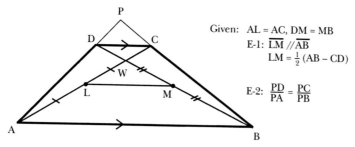

Given: AL = AC, DM = MB
E-1: $\overline{LM} \, // \, \overline{AB}$
$LM = \frac{1}{2}(AB - CD)$

E-2: $\dfrac{PD}{PA} = \dfrac{PC}{PB}$

**Figure 4.5** Properties of trapezoids defined exclusively.

## TRAPEZOIDS ON THE COORDINATE PLANE

Most geometry textbooks today have sections devoted to the placement of polygons in convenient locations on the coordinate plane. When a trapezoid has bases $b_1$ and $b_2$ and height $h$, then a coordinate system can be located so that the vertices of the trapezoid are $(0, 0)$, $(b_1, 0)$, $(a, h)$ and $(a + b_2, h)$, as shown in Figure 4.6.

If $b_1 = b_2$, then the quadrilateral has one pair of opposite sides that are parallel and equal in length and the quadrilateral can be proved to be a parallelogram. Letting the base of the parallelogram be $b$, its vertices become $(0, 0)$, $(b, 0)$, $(a, h)$, and $(a + b, h)$. This is a convenient location for a parallelogram on the coordinate plane. Thus here, as was true with the formula for the area of a trapezoid, the parallelogram is a special case of the trapezoid and the inclusive definition is the more natural one to use.

An important application of trapezoids is its use in the trapezoidal rule in calculus in which areas of quadrilaterals with vertices $(a, 0)$, $(a + n, 0)$, $(a + n, f(a + n))$, and $(a, f(a))$ are added to estimate the area under the graph of the function $f$. When $f(a + n) = f(n)$, then the quadrilateral is not only a parallelogram but also a rectangle. So if a student knows only the exclusive definition of trapezoid, then some of the figures whose areas are added in the trapezoidal rule are not trapezoids and it would seem that the rule is misnamed.

The preponderance of advantages to the inclusive definition of trapezoid has caused all the articles we could find on the subject, and most college-level geometry books, to favor the inclusive definition. The inclusive definition is also the virtual unanimous choice of geometers and other mathematicians, judging from opinions expressed in the chat-room records of the Math Forum and in essays listed on the internet (Whitely 2002 and Math Forum, Trapezoid definition discussion). If definitions evolve in the 21st century as they have in prior centuries, perhaps by the end of the century the inclusive definition of trapezoid will be the one used by a majority of authors.

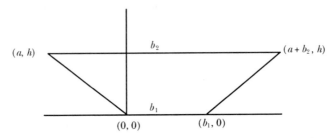

**Figure 4.6**  A convenient placement of a trapezoid on the coordinate plane.

# CHAPTER 5

# RECTANGLES

Of all the special quadrilaterals, rectangles are undoubtedly the easiest to find outside mathematics. They are the mathematical models of most ceiling panels, floor tiles, windows, sidewalk parts, sheets of paper, and the faces of most walls, doors, bricks, tables, and myriads of other objects. Rectangles are also of fundamental importance within mathematics. Depending on the point of view you wish to take regarding whether the application or the operation came first historically, the area of a rectangle is either an important application and representation of multiplication or the calculation of areas of rectangles is one of the reasons that multiplication is an important operation.

The word "rectangle" comes to us directly from the Latin "rectangulus," a combination of "rectus" *right* or *straight* and "angulus" *angle* or *small bend*. At first a rectangle referred only to a single right angle. Today's geometry textbooks are in agreement about the set of figures that are called *rectangles*. A *rectangle* universally refers to any quadrilateral with four right angles.

Words with a similar origin include: rectify, erect, regular, region, ruler, and ankle (a bend in the foot), to angle in fishing (to use a hook, something with a bend), and anchor (a weight with a hook on it). The words "England" and "English" come to us from the people called "Angles" by the Saxons who lived in today's Britain. The Angles either were so-called because they were fishermen or because they came from a part of Jutland (in the north of today's Denmark) that is in the shape of a hook.

*The Classification of Quadrilaterals: A Study of Definition*, pages 33–40
Copyright © 2008 by Information Age Publishing

The definitions of "rectangle" found in geometry textbooks possess characteristics different from those either of parallelograms or trapezoids. Unlike trapezoids, there is in recent books no dispute over which figures should be classified as rectangles. Yet, unlike parallelograms, a variety of definitions are in use. In the 86 texts we examined we found eight different defining conditions for rectangle, numbered here Re(1) through Re(8), as follows:[1]

| A rectangle is... | Number of texts |
| --- | --- |
| Re(1): a parallelogram with four right angles | 35 |
| Re(2): a parallelogram in which at least one angle is a right angle | 30 |
| Re(3): an equiangular parallelogram | 7 |
| Re(4): a quadrilateral which has four right angles | 6 |
| Re(5): an equiangular quadrilateral | 3 |
| Re(6): a rhomboid[a] with a right angle | 2 |
| Re(7): a parallelogram whose adjacent sides are unequal and whose angles are right angles | 1 |
| Re(8): a quadrilateral with three right angles | 1 |

[a] The two books, Quinn 1904 and Young 1833, that define a rectangle in terms of a rhomboid define rhomboid as a parallelogram whose adjacent sides are unequal. See Appendices A and B.

The images students have of figures are generally determined not by the words in definitions, but by pictures of the figures. Judging from the pictures of rectangles placed by their definition in today's geometry textbooks, the student reader might think that the sides of a rectangle must be horizontal and vertical, with the horizontal sides longer (for a broader look at the topic of defining ostensively[2] see, Prevost 1985). We found that 65 of the 86 books we examined with a picture by the definition showed such a rectangle. Lower van Hiele levels of understanding of geometry (see, van Hiele 1986) are associated with students who do not realize the generality of geometric figures. It would be interesting to study the extent to which the pictures of geometric figures students see in their mathematics class affect their learning of that mathematics.

## ANALYSIS OF DEFINING CONDITIONS FOR RECTANGLES

To bring some order to this array of definitions, we first sort out those that are not equivalent to each other. Re(6) and Re(7) require that a rectangle

---

1. Two books gave no definition. One book gave two definitions, both of which are counted above.
2. Robinson, Definition, pp. 117–126. Also discussed in Edwards and Ward 2004.

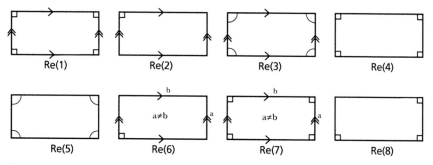

**Figure 5.1** Defining conditions for rectangles as found in geometry textbooks.

have unequal adjacent sides. This definition of rectangle excludes squares. It is noteworthy that all five books with one of these definitions were published before 1930. Thus today it is common practice not only to use an *inclusive* definition of rectangle, but also to stress how wrong it is to think of squares as not being rectangles.[3] But before 1930 it was reasonably common practice to use an *exclusive* definition of rectangles, one that does not allow squares to be rectangles. The change was certainly influenced by the report of the National Committee on Mathematical Requirements (1923, 1927). This committee consisted of representatives from the Mathematical Association of America and various mathematics teacher groups.[4] One of the many recommendations of this report was that inclusive definitions be used for the various figures in geometry. As can be seen by examining the data in Appendix B, there was significant change in definitions of the various types of quadrilaterals after this report was published.

The other six defining conditions are equivalent. They are easily seen to fall into two categories. Re(1), Re(2), and Re(3) consider a rectangle by definition to be a special type of parallelogram. Re(4), Re(5), and Re(8) consider a rectangle by definition only to be a special type of quadrilateral. This difference in classification results in slightly different proofs of equivalence of the definitions. First we consider the equivalence of those definitions in which rectangles are classified as parallelograms. Referring to Figure 5.1 may help in sorting out the various defining conditions.

---

3. See, e.g., Forbes and Eicholz, *Mathematics for Elementary Teachers*, p. 455, and Bennett and Nelson, *Mathematics for Elementary School Teachers*, p. 311.
4. The National Committee on Mathematical Requirements was first formed in 1916. Its report was the most influential report in mathematics education in the first half of the 20th century. Its bringing together key figures from mathematics teacher organizations in the East and Midwest was a prime catalyst for the formation of the National Council of Teachers of Mathematics in 1921.

**Re(1) ⟹ Re(2):** If there are four right angles, then there is at least one right angle.

**Re(2) ⟹ Re(1):** Suppose a figure ABCD is a parallelogram with a right angle at A. Then, by the definition of right angle, $\overline{AB} \perp \overline{AD}$. Now, because ABCD is a parallelogram, $\overline{AB} \,//\, \overline{CD}$. Since a line perpendicular to one of two parallel lines is perpendicular to the other, $\overline{AD} \perp \overline{CD}$. Consequently, angle D is a right angle. Repeating this same argument, we can go around the parallelogram and show that angles C and B are right angles. So ABCD has four right angles.

**Re(1) ⟹ Re(3):** This one is obvious. When a parallelogram has four right angles, then they all have the same measure and the figure is equiangular.

**Re(3) ⟹ Re(1):** Suppose a parallelogram is equiangular. To show that all the angles are right angles, we invoke the theorem that the sum of the measures of the angles of a quadrilateral is 360°. Thus each of the four angles has measure 90° and all are right angles.

This completes the proof that the three definitions Re(1), Re(2), and Re(3) are equivalent.

Now we show the equivalence of Re(4), Re(5), and Re(8).

**Re(4) ⟹ Re(5):** The proof is identical to that which showed Re(1) ⟹ Re(3).

**Re(5) ⟹ Re(4):** The proof is identical to that which showed Re(3) ⟹ Re(1). That is, to show that an equiangular quadrilateral has four right angles, we invoke the fact that the sum of the measures of the angles of the quadrilateral is 360°.

**Re(4) ⟹ Re(8):** If there are four right angles, then there are obviously three right angles.

**Re(8) ⟹ Re(4):** Given three right angles, since the sum of the measure of the angles of any quadrilateral is 360°, the remaining angle has a measure of 90°. Thus all four angles are right angles.

Now we only need to show that one of the defining conditions in the first group is equivalent to one in the second group. There are many easy ways to do this. We show that Re(1) and Re(4) are equivalent.

**Re(4) ⟹ Re(1):** Suppose a quadrilateral has four right angles. Then, because two lines perpendicular to the same line are parallel, its opposite sides are parallel. Thus the quadrilateral is a parallelogram.

**Re(1) ⟹ Re(4):** By definition, all parallelograms are quadrilaterals. So any parallelogram with four right angles is a quadrilateral with four right angles.

# WHY ARE SO MANY DIFFERENT DEFINING CONDITIONS IN USE?

A textbook author is obviously faced with a choice of which defining condition to use for rectangle.[5] We offer some thoughts regarding these choices, recognizing that other reasons for selecting a particular choice are likely. We consider only the five conditions in recent use.

Definitions Re(1), Re(2), and Re(3) help to indicate where rectangles fall in the hierarchy of types of quadrilaterals. They take advantage of the universally-accepted inclusive definition of parallelogram to allow the theorems that have been proved for parallelograms to be applied to rectangles. For instance, with these definitions, one does not have to prove that the opposite sides of a rectangle are congruent, or that the diagonals of a rectangle bisect each other (since the properties would have already been deduced for parallelograms). For this reason, in most deductive treatments of geometry, parallelograms are studied before rectangles.

Re(2) gives no more information than is necessary to show that a parallelogram is a rectangle. Re(3) gives more information than would be needed to show that a parallelogram is a rectangle; all that would be needed is that two adjacent angles of the parallelogram have the same measure. Thus it would be easier to show that a figure is a rectangle in books using Re(2) than Re(3), but it is easier to deduce properties of a rectangle from Re(3) than Re(2). Re(1), requiring four right angles, has the largest set of requirements for showing that a figure is a rectangle. But these additional conditions make Re(1) the easiest to use in deducing properties of rectangles.

Thus in choosing one definition over the others, authors are making a decision whether they want it to be easier for students to deduce that a figure is a rectangle or easier to deduce properties of rectangles once they have them. The popularity of Re(1) contrasts with the usual desire in discourse among mathematicians to include no more in the defining condition than is necessary, that is, to employ *minimal* conditions. Of the first four conditions listed above, the only conditions to appear in more than two of the texts we examined, only Re(2) is minimal.

Condition Re(4) is the one that most fits with one's intuition about rectangles and with the etymology of the word. Also, by embedding rectangles in quadrilaterals, Re(4) is more consistent with young children's

5. As with *parallelogram*, a variety of possible defining conditions for *rectangle* exist that we have not found in textbooks. Here are two: A quadrilateral is a rectangle if and only if its diagonals are congruent and bisect each other. A quadrilateral is a rectangle if and only if the perpendicular bisectors of two adjacent sides are lines of symmetry for it.

encounters with quadrilaterals, where rectangles are experienced before parallelograms.

Another condition that does not mention parallelograms, Re(5), is minimal, for a quadrilateral with three angles equal might not be a rectangle. However, because perpendicular lines are often found in the figures of elementary geometry, even though Re(4) seems to contain a more stringent criterion than Re(5) for proving that a quadrilateral is a rectangle, in practice Re(5) tends to be more difficult to apply.

Then why would Re(5) be used? It is likely because conditions Re(3) and Re(5) highlight an analogy between rectangles and rhombuses. All three of the texts using defining condition Re(5) define a rhombus as a quadrilateral in which all four sides are equal in length/congruent. Six out of the seven texts using defining condition Re(3) define a rhombus as a parallelogram with four congruent/equal sides. The remaining text using defining condition Re(3) defines a rhombus as a parallelogram whose angles are oblique and sides are equal.

## THE POSSIBLE INFLUENCE OF NON-EUCLIDEAN GEOMETRIES

In choosing a defining condition for a rectangle, it is possible that some authors have been influenced by what happens with the same condition in certain non-Euclidean geometries. In particular, there may be some influence of the hyperbolic non-Euclidean geometry of Lobatchevsky that was modeled by Poincaré and found in the *Circle Limit* prints of M. C. Escher (Dunham 2003a; Dunham 2003b; Schattschneider 1990). While it is not the intent of this study to go into non-Euclidean geometry in any detail, a summary of the features and quadrilaterals in it is interesting in how they relate to some of the conditions for a rectangle that are found in high school geometry textbooks. The reader interested in more detail might wish to consult Moise (1963) or Greenberg (1974).

We shall use the prefix "*h-*" to identify terms in this geometry. In the Poincaré-disc[6] model of hyperbolic non-Euclidean plane geometry, space (the *h-plane*) is the set of all points (the *h-points*) in the interior of a Euclidean circle Z. An *h-line* is either (1) a diameter of Z without its endpoints, or (2) the set of all points in the interior of Z that are on a Euclidean circle that intersects Z in such a way that the tangents to the two circles at the point of intersection are perpendicular. Two Euclidean circles that intersect in this way are called *orthogonal*.

---

6. We noted in writing this section that, using hits from an internet search as a metric, about twice as many people name this the Poincaré disc as name it the Poincaré disk.

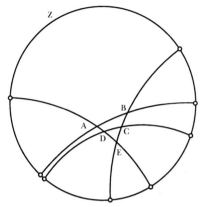

**Figure 5.2**   A quadrilateral ABCD and triangle DEC in the Poincaré-disc model of hyperbolic non-Euclidean geometry.

Figure 5.2 shows an *h*-quadrilateral ABCD and an *h*-triangle DEC in this *h*-geometry. The measure of the *h*-angle BAD is the measure of the Euclidean angle between the tangents to the two circles at the vertex A. If we extend the usual Euclidean definitions of quadrilaterals and parallel lines to their hyperbolic counterparts, then $\overline{AB} \parallel \overline{CD}$. It can be proved (see, e.g., Moise 1963) that *h*-points and *h*-lines satisfy all of the basic properties of points and lines in Euclidean geometry with the exception of the parallel postulate. Through an *h*-point, there are many *h*-lines that do not intersect a given *h*-line.[7]

As a result of this property of parallels, the sum of the measures of the angles of an *h*-triangle is less than 180°, and so the sum of the measures of the angles of an *h*-quadrilateral is less than 360°. Consequently, an *h*-quadrilateral cannot have four right angles. But there does exist an *h*-quadrilateral with three right angles; such a quadrilateral is called a *Lambert quadrilateral* after the mathematician Johann Lambert [1728–1777] who was trying to deduce as many properties as he could of Euclidean geometry without using the parallel postulate. Figure 5.3 pictures a Lambert quadrilateral in the Poincaré model.

Thus an author who wishes to extend the Euclidean defining conditions for quadrilaterals to their hyperbolic counterparts and uses Re(1), Re(4) or Re(7), that an *h*-quadrilateral with four right angles is an *h*-rectangle, would conclude that there are no *h*-rectangles. However, an author who

---

7.   A distinction is often made between *parallel h*-lines and *non-intersecting h*-lines. In Figure 5.2, the *h*-lines $\overline{AB}$ and $\overline{CD}$ are non-intersecting but not parallel. To be parallel, the Euclidean circles that contain them would have to intersect on the circle Z.

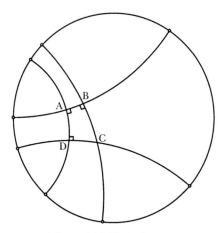

**Figure 5.3**   A Lambert quadrilateral ABCD in *h*-geometry.

extends Re(2), Re(3) Re(5), Re(6), or Re(8) to apply to *h*-quadrilaterals would conclude that there are *h*-rectangles.

This issue illustrates the subtle interplay that occurs between an intuitive idea of a figure and the mathematical definition of that figure. Textbooks and treatises on hyperbolic geometry are in unanimous agreement that there are no *h*-rectangles. This conclusion results from the view that the most important property of a rectangle is that it have four right angles, probably the most intuitive property of rectangles. A different defining condition for rectangles might lead to a different conclusion not in Euclidean geometry, but in *h*-geometry.

# CHAPTER 6

# ISOSCELES TRAPEZOIDS

The word "isosceles" is derived from the Greek "isos" *equal* and "skelos" *leg*. This word is used in English only as an adjective to describe triangles and trapezoids in which two sides (the legs) have the same length. The legs of a symmetric stick figure form an isosceles triangle with a horizontal segment connecting the person's feet. Cut that triangle off by a horizontal segment and an isosceles trapezoid is formed (Figure 6.1).

**Figure 6.1** A picture of the etymology of "isosceles."

We found four different defining conditions of *isosceles trapezoids* in use, but one definition dominates.[1]

---

1. Nineteen books gave no definition.

---

*The Classification of Quadrilaterals: A Study of Definition*, pages 41–48

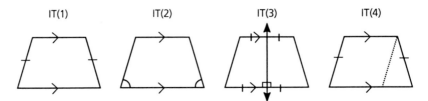

**Figure 6.2**  Defining conditions for isosceles trapezoids, as found in geometry textbooks.

| An isosceles trapezoid is... | Number of texts |
|---|---|
| IT(1):  a trapezoid in which the nonparallel sides (legs) are congruent | 64 |
| IT(2):  a trapezoid which has a pair of base angles equal in measure | 1 |
| IT(3):  a trapezoid which is symmetric about a line through the midpoints of its bases | 1 |
| IT(4):  a trapezoid in which at least one pair of opposite sides are congruent | 1 |

Under IT(1), by far the most common condition, no figure that is a parallelogram can be an isosceles trapezoid. Conditions IT(2), IT(3), and IT(4) are inclusive and are found in books that use also an inclusive definition for trapezoid. Under each of these conditions, there can exist figures that are both isosceles trapezoids and parallelograms.

## SYMMETRY AS A DEFINING CONDITION

Condition IT(3) is the only defining condition for any quadrilateral that we have seen in a geometry textbook that involves symmetry, though we did note earlier that parallelograms could be defined as quadrilaterals with rotation symmetry, and rectangles could be defined as quadrilaterals in which the perpendicular bisectors of two adjacent sides are symmetry lines. As with rectangles, the symmetry in isosceles triangles is *reflection symmetry* or *line symmetry*. If the quadrilateral is ABCD as in Figure 6.3a, and M and N are the midpoints of the parallel sides $\overline{AB}$ and $\overline{CD}$, respectively, then the symmetry means that the reflection image of ABCD over $\overline{MN}$ is the figure itself. Specifically, A and B are reflection images of each other, as are C and D. By the usual definition of reflection image, this means that $\overline{MN}$ is the perpendicular bisector of $\overline{AB}$ and of $\overline{CD}$.

If it has been assumed or proved that reflections preserve distance, then IT(3) implies IT(1) because the nonparallel sides are reflection images of each other. The preservation of distance implies the preservation of angle measure because of SSS Congruence, so IT(3) implies IT(2) since the base

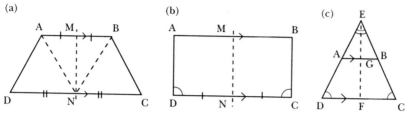

**Figure 6.3** Figures for proving equivalence of conditions IT(2) and IT(3).

angles are also reflection images of each other. If these properties of reflections have not been assumed or proved, the proofs that IT(1), IT(2), and IT(3) are equivalent conditions are longer than we have seen for other equivalent conditions, but they are very instructive in that an analysis of each argument can show how every bit of the given information needs to be employed.

**IT(3) ⇒ IT(2):** Let the midpoints of the bases $\overline{AB}$ and $\overline{CD}$ of trapezoid ABCD be M and N, respectively, as in Figure 6.3(a). Suppose ABCD is symmetric to $\overleftrightarrow{MN}$. Then $\overleftrightarrow{MN}$ is the perpendicular bisector of $\overline{AB}$. So AM = MB and m∠AMN = m∠BMN. And since MN = MN, by the SAS congruence condition, △AMN ≅ △BMN. So, as corresponding parts, m∠ANM = m∠BNM, m∠MAN = m∠MBN, and AN = BN. Now $\overleftrightarrow{MN}$ is also the perpendicular bisector of $\overleftrightarrow{CD}$, so m∠CNM = m∠DNM. Subtracting the equal angles ANM and BNM from the right angles at N, m∠DNA = m∠CNB. And since DN = CN, △AND ≅ △BNC by the SAS congruence condition. Consequently, m∠D = m∠C. Thus a pair of base angles of ABCD are equal in measure.

**IT(2) ⇒ IT(3):** Suppose that the base angles C and D of the trapezoid ABCD are equal in measure. Now consider the lines containing the sides $\overline{AD}$ and $\overline{BC}$. These sides are not the bases. Either these sides are parallel (Figure 6.3b) or they intersect at some point E (Figure 6.3c). (The point E may be such that A is between E and D, or such that D is between E and A, but the proof is essentially the same either way.)

If $\overline{AD}$ // $\overline{BC}$ (Figure 6.3b), then ABCD is a parallelogram. Furthermore, angles C and D, being on the same side of the transversal $\overleftrightarrow{CD}$, are supplementary, and since they are given to be equal in measure, they must be right angles. So $\overleftrightarrow{AD} \perp \overleftrightarrow{CD}$ and $\overleftrightarrow{BC} \perp \overleftrightarrow{CD}$. Now let N be the midpoint of $\overleftrightarrow{CD}$ and let the perpendicular to $\overleftrightarrow{CD}$ through N intersect $\overleftrightarrow{AB}$ at M. Then C and D are reflection images of each other over the line $\overleftrightarrow{MN}$. It remains to show that A and B are reflection images of each other over $\overleftrightarrow{MN}$. $\overleftrightarrow{MN} \perp \overleftrightarrow{AB}$ because a line perpendicular to one of two parallel lines is perpendicular to the other. And $\overleftrightarrow{MN}$ // $\overleftrightarrow{AD}$ and $\overleftrightarrow{MN}$ // $\overleftrightarrow{BC}$ because two lines perpendicular to the same line are parallel. Consequently, both AMND and BMNC are parallelograms. AM = DN because opposite sides of a parallelogram have the same measure, and likewise BM = CN. Now, recalling that N was the midpoint of $\overline{CD}$, CN = DN, so AM = BM, and M is the midpoint of $\overline{AB}$. Consequently, A and B are reflection images of each other over $\overleftrightarrow{MN}$.

Should $\overleftrightarrow{AD}$ and $\overleftrightarrow{BC}$ intersect at E (Figure 6.3c), let F be the intersection of the bisector of ∠DEC and $\overleftrightarrow{CD}$. Since $\overleftrightarrow{EF}$ is an angle bisector, m∠DEF = m∠CEF. Also, EF = EF. So ΔDEF ≅ ΔCEF by the AAS congruence condition. Since DF and CF are corresponding sides of these triangles, DF = CF. And since angles DFE and CFE are corresponding angles of these triangles, m∠DFE = m∠CFE and they must be right angles. Thus $\overleftrightarrow{EF}$ is the perpendicular bisector of $\overline{CD}$.

We now need to show that $\overleftrightarrow{EF}$ is the perpendicular bisector of $\overline{AB}$. Since $\overline{AB}$ and $\overline{CD}$ are the bases of trapezoid ABCD, $\overline{AB}$ // $\overline{CD}$. Consequently, as corresponding angles formed by transversals to the parallel lines, m∠BAE = m∠CDE and m∠ABE = m∠DCE. Hence m∠ABE = m∠BAE. Now, let G be the intersection of $\overleftrightarrow{EF}$ and $\overleftrightarrow{AB}$. Following the argument used earlier, ΔAGE ≅ ΔBGE by the AAS condition. Thus $\overleftrightarrow{EF}$ is the perpendicular bisector of $\overline{AB}$. As a consequence, A and B are reflection images of each other over the line $\overleftrightarrow{EF}$, as are C and D. This means that $\overleftrightarrow{EF}$ is a line of symmetry for the figure.

Notice that in Figure 6.3b, the isosceles trapezoid is a rectangle. This is only possible under the inclusive definition of isosceles trapezoid that is in force when conditions IT(2) or IT(3) are being used. When condition IT(1) is being used, the situation of Figure 6.3b is not allowed because the non-base sides cannot be parallel. However, the arguments above do show that when IT(1) is being used, conditions IT(2) and IT(3) are equivalent. Now we show that they are then equivalent also to IT(1).

**IT(3) ⇒ IT(1):** Use what has been deduced in the proof of IT(3) ⇒ IT(2) above. Since ΔAND ≅ ΔBNC, AD = BC and the two sides not given as the bases have the same length. Thus IT(3) ⇒ IT(1).

**IT(1) ⇒ IT(3):** Let trapezoid ABCD be given with bases $\overline{AB}$ and $\overline{CD}$ and suppose AD = BC as in Figure 6.4. Since the condition IT(1) is given, we know that $\overleftrightarrow{AD}$ and $\overleftrightarrow{BC}$ are not parallel. It is also the case that neither of these lines is perpendicular to $\overline{AB}$, for if AD ⊥ $\overline{AB}$, then also AD ⊥ $\overline{CD}$ and AD is the shortest distance between the bases. If that is so, then BC must also be that shortest distance and so BC ⊥ $\overline{AB}$. Then both $\overleftrightarrow{AD}$ and $\overleftrightarrow{BC}$ would be perpendicular to the same line, so they would be parallel, contradicting IT(1). Now let G be the foot of the perpendicular from A to $\overleftrightarrow{CD}$, and let H be the foot of the perpendicular from B to $\overleftrightarrow{CD}$. AG and BH are parallel because they are perpendicular to the same line, and thus AGHB is a parallelogram, and as opposite sides, AG = BH. Now ΔAGD ≅ ΔCHB by the HL condition. Angles C and D, being corresponding angles in these congruent triangles, are thus equal in measure.

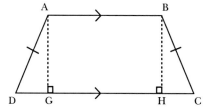

**Figure 6.4** Proving that base angles of an isosceles trapezoid have the same measure.

Since conditions IT(1) and IT(3) are equivalent for isosceles trapezoids defined exclusively, and conditions IT(2) and IT(3) are also equivalent then, the three conditions are equivalent for these isosceles trapezoids.

Condition IT(4) merits a special comment. It is found in the Moise and Downs *Geometry* (1964) whose first author Edwin E. Moise, a topologist, had written a widely used book on the foundations of geometry[2] and was the main author of the School Mathematics Study Group *Geometry*. In the 1964 edition, trapezoid is defined in the exposition as a quadrilateral with at least one pair of opposite sides parallel, and isosceles trapezoid in a problem as "a trapezoid in which at least one pair of opposite sides are congruent." This is the most inclusive pair of definitions we have seen, and it is clearly not meant to be an error, for the book specifically notes that under these definitions all parallelograms are *isosceles* trapezoids. However, in a later edition (1971), the book defines trapezoid in an exclusive manner, and in the corresponding problem defines isosceles trapezoid using definition (1) on this page.

Thus, even though at first glance it might seem that condition IT(4), being analogous to IT(1), is the most natural of the three inclusive conditions, IT(4) presents major difficulties. Few people want all parallelograms to be isosceles trapezoids, which would then also classify all rhombuses, rectangles, and squares as isosceles trapezoids. It is precisely to avoid the quite inclusive defining condition IT(4) that condition IT(2) or IT(3) is employed instead.

## SHOULD RECTANGLES BE ISOSCELES TRAPEZOIDS?

As with the relationship between parallelograms and trapezoids, we ask whether it is helpful for rectangles to be defined so that they are isosceles trapezoids and we study the question by considering theorems about isosceles trapezoids that are defined exclusively. Do these theorems hold for rectangles? The first theorems that come to mind are exactly those of the defining conditions IT(2) and IT(3). We identify a theorem here as IT-E

---

2. Moise, *Elementary Geometry from an Advanced Standpoint*, 1963.

if it is true only when trapezoids are defined exclusively, and we identify a theorem as IT-I if the theorem is true under an inclusive definition of trapezoids.

**Theorem (IT-I1):** In an isosceles trapezoid, each pair of base angles is equal in measure.

**Theorem (IT-I2):** The line containing the midpoints of the bases of an isosceles trapezoid is a line of symmetry for the trapezoid.

Any property of all isosceles trapezoids defined inclusively will also be true for all isosceles trapezoids defined exclusively because the first group of isosceles trapezoids contains the second. Thus Theorems IT-I1 and IT-I2 are true for rectangles even if isosceles trapezoids are not defined to include rectangles. Here is another theorem with this property.

**Theorem (IT-I3):** The diagonals of an isosceles trapezoid have the same length.

Still another theorem true under either an exclusive or an inclusive definition states a condition that could serve as a definition of *isosceles trapezoid.*

**Theorem (IT-I4):** *If* the diagonals of a trapezoid have the same length, then the trapezoid is an isosceles trapezoid.

When there is an inclusive definition of trapezoid, Theorem IT-I4 allows the given trapezoid to be a parallelogram. In this case, the isosceles trapezoid is a rectangle.

In the coordinate plane, a convenient placement of an isosceles trapezoid has the $x$-axis as one base, this yields the consecutive vertices $(a, h)$, $(b - a, h)$, $(b, 0)$, and $(0, 0)$, as shown in Figure 11.10. If $a = 0$, then this figure is a rectangle. So again, there is a natural way in which isosceles trapezoids include rectangles.

But some theorems under the exclusive definition are not true statements under the inclusive definition. One of these is the following theorem that we have mentioned earlier, which relates isosceles trapezoids to isosceles triangles.

**Theorem (IT-E1):** Suppose $\overline{AB}$ and $\overline{CD}$ are the bases of an isosceles trapezoid ABCD. Then the lines $\overleftrightarrow{AD}$ and $\overleftrightarrow{BC}$ intersect at a point, call it E, and triangles EAB and ECD are isosceles triangles.

This theorem is basic enough to cause some authors to select the exclusive definition of trapezoid. Authors who select the inclusive definition would modify the theorem to something like the following:

**Theorem (IT-I5):** Suppose $\overline{AB}$ and $\overline{CD}$ are the bases of an isosceles trapezoid ABCD. Then, *if* the lines $\overleftrightarrow{AD}$ and $\overleftrightarrow{BC}$ intersect at a point E, then triangles EAB and ECD are isosceles triangles.

Two other theorems that are true only under the exclusive definition of an isosceles trapezoid are the following:

**Theorem (IT-E2):** The bases of an isosceles trapezoid have different lengths.

**Theorem (IT-E3):** An isosceles trapezoid has exactly one line of symmetry.

Each can be modified so that it is true under the inclusive definition.

**Theorem (IT-I6):** If bases of an isosceles trapezoid have the same length, then the figure is a rectangle.

**Theorem (IT-I7):** An isosceles trapezoid has at least one line of symmetry; if it has two, then it is a rectangle.

The upshot of this analysis is that if parallelogram is defined so that every parallelogram is a trapezoid, then every rectangle is not only a trapezoid, but an isosceles trapezoid as well.

# CHAPTER 7

# KITES

A special quadrilateral called a *kite* is defined in only a minority (18 of 86) of all the textbooks we examined, but we found this type in a majority of the geometry textbooks published since 1990 and absent from only two texts published since 2000.

The 18 geometry texts with definitions of kite offer six different defining conditions.[1]

| A kite is... | Number of texts |
|---|:---:|
| K(1): a quadrilateral which has two distinct[a] pairs of adjacent sides of the same length | 10 |
| K(2): a quadrilateral in which exactly one diagonal is a perpendicular bisector of the other | 3 |
| K(3): a quadrilateral in which one diagonal is a perpendicular bisector of the other | 2 |
| K(4): a quadrilateral with two pairs of adjacent sides congruent and no opposite sides congruent | 1 |
| K(5): a convex quadrilateral with two distinct pairs of adjacent sides of the same length | 1 |
| K(6): a four-side with an axis of symmetry | 1 |

[a] The words "disjoint" or "non-overlapping" are often used in place of "distinct." These modifiers for the word "pair" exist to avoid the possibility that {a, b} and {a, c} are considered as two different pairs.

1. Two texts, Fischer and Hayden 1965 and Nichols et al. 1974, give defining conditions that were grouped with K(1) although the wording varied slightly. See Appendices A and B.

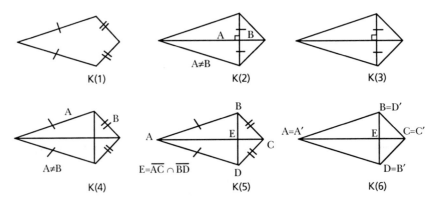

**Figure 7.1**   Kites.

Five different sets of figures are determined by the six conditions above! Only definitions K(1) and K(4) are equivalent. Definitions K(4) and K(2) do not allow all four sides of a kite to have the same length, so under these definitions no rhombuses are kites. Definitions K(2) and K(5) require that a kite be convex. (If the diagonals do not intersect, then while the line containing one can bisect and be perpendicular to the other, it cannot go both ways, as seen in the non-convex quadrilateral ABCD in Figure 7.2b.) Definition K(6) as written allows isosceles trapezoids, since isosceles trapezoids have symmetry lines, but in both books using K(6), in the figure accompanying the definition the axis of symmetry contains a vertex of the quadrilateral. No book we examined published in the last 100 years has used this definition.

Figure 7.2 shows the non-equivalence of these definitions pictorially.

Of all the names for quadrilaterals, the word "kite" is the only one whose origin is neither from Greek or Latin. Its etymology is from the Old English "*cyta*" *owl*. But its choice in mathematics is derived not from the owl, but from the child's toy at the end of a string. In German, the word "flieger" *flyer* refers both to this toy and to the geometric figures we call kites.

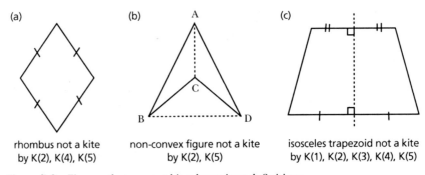

**Figure 7.2**   Figures that are not kites by various definitions.

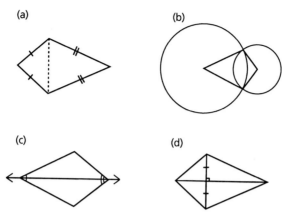

**Figure 7.3** Mathematical origins of kites.

Kites arise in at least four common geometrical contexts, shown in Figure 7.3. One is the idea of putting together two different isosceles triangles with the same base and then removing the base (Figure 7.3a). Second is the figure formed by the centers and points of intersection of two intersecting circles (Figure 7.3b). Third, in analogy with the isosceles trapezoid being a quadrilateral with a line of symmetry that is a perpendicular bisector of a side, the kite is a quadrilateral with a line of symmetry that is a bisector of an angle (Figure 7.3c). This idea is closely related to the figure formed when a triangle is reflected over one of its sides with a quadrilateral being formed by the triangle and its image. Fourth, a kite stems from considering a segment on the perpendicular bisector of a given segment and then connecting the endpoints of the two segments (Figure 7.3d).

Since all of these origins can accommodate non-convex kites, and all could allow for kites in which all four sides have the same length, the selection of definition K(5) seems to be based more on the most common shape of the child's toy and not on the mathematical origins of the idea. Definitions K(2) and K(3) may also have been selected with the idea of a child's kite in mind, for the most common toy kites are almost always made rigid by perpendicular pieces of wood that connect its opposite points.

Under any of the definitions but K(6), all kites possess the following properties:

**Theorem K-1:** In a kite, there exists a pair of opposite angles with the same measure.

**Theorem K-2:** In a kite, one diagonal bisects a pair of opposite angles. The line containing this diagonal is a symmetry line for the kite.

**Theorem K-3:** In a kite, the line containing a symmetry diagonal is the perpendicular bisector of the other diagonal.

**Theorem K-4:** The area of a kite is half the product of the lengths of its diagonals.

None of these theorems holds for all kites under definition K(6).

## EQUIVALENT DEFINING CONDITIONS FOR KITES

We have noted that the defining conditions K(1) and K(3) are equivalent to each other. Here is a proof.

**K(1) ⇒ K(3):** Suppose ABCD is a kite with AB = AD and CD = CB and let E be the intersection of the lines containing the diagonals $\overleftrightarrow{AC}$ and $\overline{BD}$. There are essentially two different figures depending on whether the kite is convex or nonconvex (Figure 7.4), but the proof is the same. Since AC = AC, △ABC ≅ △ADC by SSS congruence. As a result, the corresponding angles BAE and DAE are equal in measure. So △ADE ≅ △ABE by SAS congruence. Thus DE = BE and the corresponding angles AED and AEB are equal in measure, so $\overline{AE} \perp \overline{BD}$. Consequently, $\overleftrightarrow{AC}$ is the perpendicular bisector of $\overline{BD}$.

**K(3) ⇒ K(1):** Suppose ABCD is a kite whose diagonals intersect at E and suppose $\overleftrightarrow{AC}$ is the perpendicular bisector of $\overline{BD}$. Again convex and nonconvex kites yield different figures but the proof is the same. Then BE = ED and angles AEB and AED are right angles. Consequently, △AEB ≅ △AED by SAS congruence. Then AB = AD. Also, △BEC ≅ △DEC by SAS congruence, so BC = DC. So the kite has two distinct pairs of consecutive sides of the same length.

## ISOSCELES TRIANGLES AS DEGENERATE KITES

Suppose A, B, C, and D are distinct points, AB = AD, and CB = CD. Then on most occasions a quadrilateral ABCD is formed. It will look like one or the other of the situations of Figure 7.4. But if one of A or C is the midpoint of $\overline{BD}$, then an isosceles triangle is formed. If C is the midpoint of $\overline{BD}$, then C is also the midpoint of the base of the isosceles triangle ABD. We can think of this isosceles triangle as a *degenerate kite*. The diagonals of the kite are still $\overline{AC}$ (the altitude of the isosceles triangle) and $\overline{BD}$ (the base of the isosceles triangle).

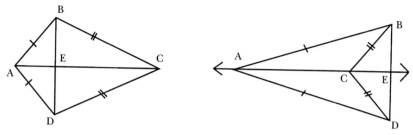

**Figure 7.4** Convex and nonconvex kites ABCD.

Theorems K-1 through K-4 described above now become theorems about isosceles triangles.

**Theorem K-1:** In a degenerate kite (isosceles triangle), there exists a pair of opposite angles with the same measure.

**Theorem K-2:** In a degenerate kite, one diagonal bisects a pair of the opposite angles. The line containing this diagonal is a symmetry line for the degenerate kite.

**Theorem K-3:** In a degenerate kite, the line containing a symmetry diagonal (the altitude of the isosceles triangle) is the perpendicular bisector of the other diagonal (the base of the isosceles triangle).

**Theorem K-4:** The area of a degenerate kite is half the product of the lengths of its diagonals (i.e., half the product of its base and its altitude).

Thus there is a sense in which every isosceles triangle can be thought of as a degenerate kite. But it would require a more inclusive definition of kite than found in any existing book for these degenerate kites to be classified as kites. However, as we noted in Chapter 2, many definitions of polygon are inclusive enough to allow degenerate quadrilaterals of this type.

# CHAPTER 8

# RHOMBUSES (RHOMBI)

As with the kite, the origin of the term "rhombus" is a child's toy. The Greek word "rhombos" ρομβοσ, from "rhomb" *turn*, referred to an object called a bull-roarer that was attached to the end of a cord that was rapidly twirled around. It received its name because the twirling caused a roaring sound and the object was used by the ancient Minoans to taunt bulls. The faces of the bull-roarer were quadrilaterals with sides of equal length, so the name rhombos was attached to this type of quadrilateral.[1] The word "rhomb," used in Great Britain for "rhombus," is not found in recent geometry text books published in the U.S.

In the 86 high school texts examined we found seven different defining conditions for *rhombus*, as follows.[2]

| A rhombus is... | | Number of texts |
|---|---|---|
| Rh(I1): | a parallelogram with four equal sides | 37 |
| Rh(I2): | a parallelogram in which at least two consecutive sides are congruent | 20 |
| Rh(I3): | a quadrilateral in which all four sides are equal in length/congruent | 12 |
| Rh(E1): | a rhomboid having all sides equal | 7 |
| Rh(E2): | a parallelogram whose angles are oblique and sides are equal | 6 |
| Rh(E3): | a rhomboid having two adjacent sides equal[a] | 2 |
| Rh(E4): | a parallelogram with oblique angles with two adjacent sides equal | 1 |

[a]  Both books that define a rhombus as a special kind of rhomboid have earlier defined a rhomboid as a parallelogram having oblique angles.

1.  http://www.pballew.net/rhomb.html.
2.  One book gave no definition.

*The Classification of Quadrilaterals: A Study of Definition*, pages 55–58
Copyright © 2008 by Information Age Publishing

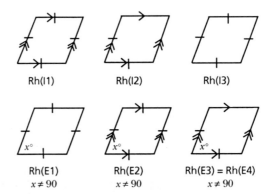

**Figure 8.1** Defining conditions for rhombuses.

The letters E and I before the number of the defining condition indicate whether the definition is exclusive or inclusive, i.e., whether defining "rhombus" in this way excludes or includes squares. A separation of the texts by publication date shows that the definition of rhombus has undergone an evolution in this regard. In the 22 books examined that were published before 1925, all seven of these conditions are found. In the 66 books published from 1925 on, except for Sigley and Stratton (1948) and Hart (1959), only the inclusive conditions are found. The major change that occurred in 1925 is again likely due to the publication of the recommendations of the National Committee on Mathematical Requirements, in which an inclusive definition for all figures is recommended.[3]

The most common definition Rh(I1) contains more information than is needed, while both Rh(I2) and Rh(I3) contain minimal information. For rhombuses, the desire for being able to easily deduce properties of rhombuses seems about as important as the desire to make it easy to deduce that a figure is a rhombus.

Rh(I1) implies Rh(I2) obviously; if four sides are equal in length, then two adjacent sides must be equal. Rh(I2) implies Rh(I1) because opposite sides of a parallelogram are congruent. So Rh(I1) and Rh(I2) are equivalent.

Rh(I1) also obviously implies Rh(I3). Conversely, since a quadrilateral with both pairs of opposite sides equal is a parallelogram, if a quadrilateral has four equal sides, then it satisfies Rh(I1). So Rh(I3) and Rh(I1) are equivalent.

Essentially the same arguments can be used to prove the equivalence of Rh(E1) through Rh(E4).

A synonym for "rhombus" is *diamond*. But often the word diamond is restricted to refer to a rhombus whose diagonals lie on horizontal and

3. The National Committee on Mathematical Requirements, *The Reorganization of Mathematics in Secondary Education*, 108-09.

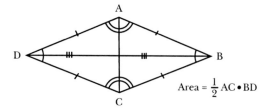

**Figure 8.2**   Properties of all rhombuses.

vertical lines. This is how the figure formed by connecting home plate and the three bases in order looks as seen from home plate and is the origin of the "baseball diamond." That is, although a baseball diamond is a square in shape, it is called a diamond because of its orientation. In the website *mathworld,*[4] a diamond is restricted to referring to a square with horizontal and vertical diagonals.

The rhombus shape is called a *lozenge* in the parlance of U.S. military insignias, heraldry, and pattern design. The *American Heritage Dictionary* views lozenge and diamond as synonymous with rhomboid. Mathworld defines a lozenge as a rhombus with acute angles (i.e., an exclusively-defined rhombus), or, more specifically, a rhombus with two 45° angles, but its use in design allows all angles.

The following are commonly taught properties of all rhombuses, regardless of the defining condition used (see Figure 8.2).

**Theorem Rh(1):**   Both pairs of opposite sides of a rhombus are parallel.

**Theorem Rh(2):**   In a rhombus, opposite angles have the same measure.

**Theorem Rh(3):**   In a rhombus, each diagonal lies on the perpendicular bisector of the other diagonal.

**Theorem Rh(4):**   In a rhombus, the line containing each diagonal bisects both angles of the rhombus at the endpoints of the diagonal.

**Theorem Rh(5):**   The area of a rhombus is one-half the product of the length of its diagonals.

**Theorem Rh(6):**   Every rhombus is symmetric to the lines containing its diagonals.

4.   Wolfram Research, http://mathworld.wolfram.com, created and maintained by Eric Weisstein.

The first two of these theorems follow immediately from the definition of "rhombus" except when condition Rh(I3) is used. The other four theorems classify a rhombus as a special kind of kite, provided an inclusive definition of kite is being used. It is noteworthy that many of the books that discuss rhombuses and also mention theorems Rh(3) through Rh(5) do not mention kites. Nor does any book we examined define a rhombus as a special kind of kite. Yet the preponderance of properties of all rhombuses that are also properties of all kites suggests that rhombuses are naturally first seen as special kites rather than special parallelograms.

If inclusive definitions are used both for kite and rhombus, then rhombuses occupy a rather central position in the quadrilateral hierarchy, as seen in Figure 8.3.

The definitions of "rhombus" and "rectangle" found most often in textbooks are analogous. Specifically, definitions Rh(1), Rh(2), and Rh(3) for the rhombus are analogous to definitions Re(1), Re(2), and Re(4) for the rectangle. In each case, equal side lengths (for the rhombus) correspond to equal angle measures (for the rectangle). When inclusive definitions are used, an analogy with rectangles holds for the hierarchy in Figure 8.3 as well. Replace kite by isosceles trapezoid and rhombus by rectangle. That is, under inclusive definitions, just as a rhombus is a figure that is both a parallelogram and a kite, a rectangle is a figure that is both a parallelogram and an isosceles trapezoid.

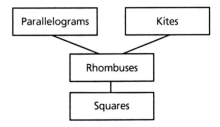

**Figure 8.3**   Rhombuses in the inclusive hierarchy.

# CHAPTER 9

# SQUARES

Geometry textbooks contain quite a number of different definitions of *square* but all the definitions we found yield the same set of figures: those quadrilaterals with four sides of the same length and four right angles. The variety of possible definitions results because, under most classifications, squares lie at the bottom of the hierarchy of quadrilaterals: under inclusive definitions, all squares are rhombuses, rectangles, parallelograms, kites, isosceles trapezoids, and trapezoids. Lying at the bottom of the food chain, so to speak, squares may be swallowed up under any of these other types of figures, or directly as a special type of quadrilateral. In the 86 high school texts examined, we found two different definitions of square as a special type of each of rectangle, rhombus, and parallelogram, one as a special kind of quadrilateral, and two as a figure that is both a rectangle and a rhombus, as follows.[1]

| A square is... | Number of texts |
|---|---|
| S(1): a rectangle with four congruent sides | 38 |
| S(2): a rectangle with a pair of consecutive congruent sides | 14 |
| S(3): a parallelogram that is both a rectangle and a rhombus | 10 |
| S(4): a parallelogram with four equal sides and four right angles | 7 |
| S(5): a parallelogram with one right angle and two adjacent sides congruent | 5 |
| S(6): a quadrilateral that has four equal sides and four right angles | 4 |
| S(7): a rhombus with four equal angles | 3 |
| S(8): a rhombus with a right angle | 2 |
| S(9): a quadrilateral that is both a rhombus and a rectangle | 1 |

1. Three books gave no definition. Three books gave two definitions.

*The Classification of Quadrilaterals: A Study of Definition*, pages 59–62
Copyright © 2008 by Information Age Publishing

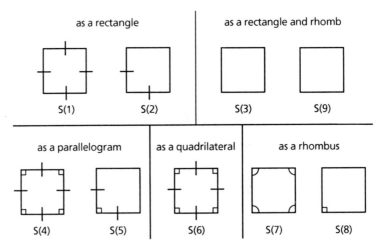

**Figure 9.1**  Defining conditions for squares.

## DEFINING QUADRILATERALS USING ONLY THE HIERARCHY

Of all the definitions of quadrilaterals in geometry textbooks, only the square has been defined as an intersection of two (or more) types of quadrilaterals. By defining a square as a figure that is both a rectangle and a rhombus, the author is emphasizing that many of the properties of squares can be traced back to properties of these other two figures.

| Because a square is a rectangle: | Because a square is a rhombus: |
|---|---|
| all its angles have the same measure | all its sides have the same length |
| its diagonals have the same length | its diagonals are perpendicular |
| it is symmetric to the perpendicular bisectors of its sides | it is symmetric to the bisectors of its angles |
| its area is the product of the lengths of its sides. | its area is half the product of the lengths of its diagonals. |

When inclusive definitions are given, it is possible to define other types of quadrilaterals in this manner. For instance, rectangles are figures that are both parallelograms and isosceles trapezoids.

| **Because a rectangle is a parallelogram:** | **Because a rectangle is an isosceles trapezoid:** |
| --- | --- |
| it has two pairs of opposite angles equal in measure | it has two disjoint pairs of adjacent angles equal in measure |
| its diagonals bisect each other | its diagonals are of equal length |
| it has 180° rotation symmetry. | it is symmetric to a line that is the perpendicular bisector of a side. |

In an analogous fashion, rhombuses are figures that are both parallelograms and kites.

| **Because a rhombus is a parallelogram:** | **Because a rhombus is a kite:** |
| --- | --- |
| it has two pairs of adjacent supplementary angles | it has two disjoint pairs of adjacent sides equal in measure |
| its diagonals bisect each other | its diagonals are perpendicular |
| it has 180° rotation symmetry. | it is symmetric to a line containing a diagonal. |

These relationships among these types of quadrilaterals strike us as elegant and beautiful. Figure 9.2 exhibits the wonderful symmetry of the inclusive hierarchy of these six types of quadrilaterals.

Does this mean that trapezoids somehow do not belong with the other special types of quadrilaterals? Not at all. It is possible to extend the pattern of the hierarchy of Figure 9.2 up to include trapezoids by introducing the figure known as a *cyclic quadrilateral.*

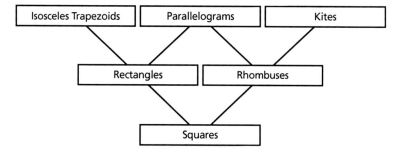

**Figure 9.2**   An inclusive quadrilateral hierarchy.

# CHAPTER 10

# CYCLIC QUADRILATERALS

A major type of quadrilateral studied by geometers but found only occasionally in pre-college textbooks is the *cyclic quadrilateral*. The cyclic quadrilateral receives its name from the property that there exists a circle that contains all four of its vertices. The word cyclic is from the Greek "kuklos" κυκλος *circle* or *wheel*. Words that share this origin convey the idea of a circle or of going around. They include: cycle, bicycle (two wheels), tricycle, unicycle, cycloid, cyclone, circulate (move around), circumference, circumstance (the situation that is around), and circus (events like those in today's circuses used to be held in a circular arena).

From its defining property, it can be proved that all isosceles trapezoids (unless defined using definition IT-4) and all rectangles are cyclic quadrilaterals. Kites are cyclic quadrilaterals if their congruent angles are also right angles. This makes the symmetry diagonal of the kite also a diameter of a circle that contains the kite's vertices.

In the 86 geometry texts we examined, 12 defined "cyclic quadrilateral." We found two different defining conditions.

| A cyclic quadrilateral is... | Number of texts |
|---|---|
| CQ(1): a quadrilateral whose four vertices lie on a circle | 7 |
| CQ(2): a quadrilateral inscribed in a circle with opposite angles supplementary | 5 |

*The Classification of Quadrilaterals: A Study of Definition*, pages 63–68
Copyright © 2008 by Information Age Publishing

**63**

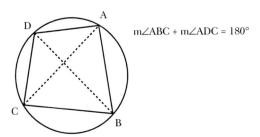

$$m\angle ABC + m\angle ADC = 180°$$

**Figure 10.1** A cyclic quadrilateral.

Ten other books hinted at a definition in the text and followed the hint with exercises that continued the defining process.[1]

The defining conditions CQ(1) and CQ(2) are equivalent. Here is a proof.

**CQ(1) ⇒ CQ(2):** Let the cyclic quadrilateral be ABCD as in Figure 10.1. Then

$$m\angle A = \tfrac{1}{2} m\,DCB \text{ and } m\angle C = \tfrac{1}{2} m\,DAB$$

because both angles are inscribed angles. So

$$m\angle A + m\angle C = \tfrac{1}{2}(m\,DCB + m\,DAB)$$

$$= \tfrac{1}{2} \cdot 360°$$

$$= 180°.$$

**CQ(2) ⇒ CQ(1):** Refer to Figure 10.2. We are given that ABCD is a (non-degenerate) quadrilateral with m∠D + m∠B = 180°. We wish to show that there is a circle containing all four points A, B, C, and D.

Since the three vertices A, B, and C are non-collinear, there is a circle containing them. Suppose the fourth vertex D is in the circle's interior (Figure 10.2a). Then $\overline{CD}$ can be extended through D to intersect the circle at D′. Because we have already proved CQ(1) ⇒ CQ(2), we know that the opposite angles of the quadrilateral ABCD′ are supplementary. So m∠D′ + m∠B = 180°. But we are given m∠ADC + m∠B = 180°. So m∠ADC = m∠D′. This is impossible because the measure of an exterior angle of a triangle (here ΔADD′) must be greater than the measure of either remote interior angle. Since the supposition has led to a contradiction, D cannot be in the circle's interior. Suppose D is in the exterior of the circle. Then let D″ be the intersection of $\overline{CD}$ and the circle (Figure 10.2b). Now

---

1. See Appendix B, where these books are listed as definition 3.

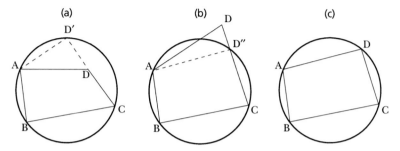

**Figure 10.2**  Proving equivalence of definitions of cyclic quadrilaterals.

we reason in a way similar to that when D is in the interior. Because D″ is on the circle, m∠AD″C + m∠B = 180°. But we are given m∠D + m∠B = 180°. So m∠AD″C = m∠D. This is impossible because ∠AD″C is an exterior angle of ΔADD″ and m∠D is a remote interior angle. Again the supposition has led to a contradiction, and so D cannot be in the exterior of the circle. So if the opposite angles B and D of quadrilateral ABCD are supplementary, the only possibility is that D is on the circle (Figure 10.2c). This ends the proof that CQ(2) ⇒ CQ(1) and shows that these defining conditions are equivalent.

Because the perpendicular bisector of a chord of a circle contains the center of the circle, the perpendicular bisectors of the four sides of a cyclic quadrilateral are concurrent at the center of the circle. There may be two, three, or four distinct lines. If there are exactly two distinct lines, then the perpendicular bisectors of both opposite sides of the cyclic quadrilateral must coincide, and so the cyclic quadrilateral is a rectangle. If there are exactly three distinct lines, then the perpendicular bisectors of exactly one pair of opposite sides coincide, and the quadrilateral is an isosceles trapezoid but not a rectangle.

Conversely, if all the perpendicular bisectors of the sides of a quadrilateral have exactly one point in common (whether there are two, three, or four distinct lines), then the quadrilateral's vertices are all equidistant from the point of concurrency, and the quadrilateral must be cyclic. This proves:

> **Theorem CQ1:**  A quadrilateral is cyclic if and only if the perpendicular bisectors of all its sides are concurrent.

All cyclic quadrilaterals possess two additional important properties. The first is due to the 2nd century Greek mathematician Claudius Ptolemy and the second is due to the 7th century Indian mathematician Brahmagupta. Neither theorem is easy to prove, which probably accounts for the lack of attention given to them and to cyclic quadrilaterals in pre-college texts.

**Theorem CQ2:** (Ptolemy's Theorem). In the cyclic quadrilateral ABCD, AC · BD = AB · CD + AD · BC. (In a cyclic quadrilateral, the product of the lengths of the diagonals equals the sum of the products of the lengths of its opposite sides.)

**Theorem CQ3:** (Brahmagupta's Theorem). The area of a cyclic quadrilateral with sides $a$, $b$, $c$, and $d$, and semiperimeter $s$ is $\sqrt{(s-a)(s-b)(s-c)(s-d)}$.

Proofs of Theorem CQ2 can be found in a number of sources.[2] The cited sources also deduce the converse, thus showing that the formula in Ptolemy's Theorem could be a defining condition for a cyclic quadrilateral. A proof of Theorem CQ3 using trigonometry can be found in Coxeter and Greitzer.[3] More generally, the area of any quadrilateral ABCD is

$$\sqrt{(s-a)(s-b)(s-c)(s-d)-abc\cos\left(\frac{A+C}{2}\right)}.\text{[4]}$$

Since the cosine in the formula is 0 exactly when A and C are supplementary angles, Brahmagupta's formula is true only for cyclic quadrilaterals. So Brahmagupta's formula also could be a defining condition for a cyclic quadrilateral.

Brahmagupta's Theorem is a generalization of the better-known Hero's (or Heron's) formula for the area of a triangle that is often found in high school geometry texts. Let $d = 0$ in Brahmagupta's formula by bringing two vertices of the cyclic quadrilateral together and Hero's formula appears.

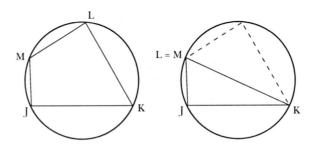

**Figure 10.3** A triangle as a cyclic quadrilateral with a side of zero length.

---

2. E.g., Johnson, *Advanced Euclidean Geometry*, pp. 62–63; Altshiller-Court, *College Geometry*, pp. 255–256; Coxeter and Greitzer, *Geometry Revisited*, p. 42.
3. Coxeter and Greitzer, op. cit., pp. 57–58.
4. See Howard Eves, *A Survey of Geometry*. Revised Edition. Boston: Allyn and Bacon, 1972, p. 49

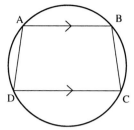

**Figure 10.4**   A cyclic trapezoid is an isosceles trapezoid.

In this regard, a triangle can be thought of as a degenerate cyclic quadrilateral.[5]

   Suppose a quadrilateral ABCD is both a cyclic quadrilateral and a trapezoid with $\overleftrightarrow{AB}$ // $\overleftrightarrow{CD}$ (Figure 10.4). Then angles A and C are supplementary because ABCD is a cyclic quadrilateral, and angles A and D are supplementary because ABCD is a trapezoid. Consequently, $m\angle D = m\angle C$. This means the trapezoid is an isosceles trapezoid, or we might say that a cyclic trapezoid is isosceles. Furthermore, every isosceles trapezoid is a cyclic quadrilateral because the opposite angles of an isosceles trapezoid are supplementary. Consequently, cyclic quadrilaterals fit into the quadrilateral hierarchy as shown in Figure 10.5.

   In the next chapter we show inclusive hierarchies of all the special types of quadrilaterals we have discussed. Notice particularly the characterization of cyclic quadrilaterals, trapezoids, parallelograms, and isosceles trapezoids using only the congruence or supplementary nature of opposite or adjacent angles. This and the other hierarchies suggest that, as quadrilaterals, cyclic quadrilaterals occupy a position quite like trapezoids and should be given greater importance in the curriculum than they currently possess.

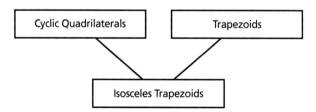

**Figure 10.5**   Cyclic quadrilaterals in the hierarchy.

---

   5.  With regard to area formulas, a triangle can also be thought of as a degenerate trapezoid in which one base has length zero. Then the formula $A = \frac{1}{2}h(b_1 + b_2)$ becomes $A = \frac{1}{2}h(b_1 + 0) = \frac{1}{2}h(b_1)$, which we write as $A = \frac{1}{2}hb$ because a triangle has only the one base.

# CHAPTER 11

# PROPERTIES OF QUADRILATERALS EXHIBITED IN HIERARCHIES

Figure 11.1 displays the inclusive quadrilateral hierarchy including cyclic quadrilaterals. It should be interpreted as follows: If a bar or a path of bars connect a figure type A to a figure of type B above it, then every figure of type A is also of type B.

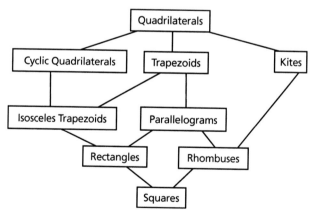

**Figure 11.1**   An inclusive quadrilateral hierarchy with eight special types of quadrilaterals.

*The Classification of Quadrilaterals: A Study of Definition*, pages 69–82
Copyright © 2008 by Information Age Publishing
All rights of reproduction in any form reserved.

## TYPES OF QUADRILATERALS AS INTERSECTIONS
## OF OTHER TYPES

A special property of the hierarchical diagram of Figure 11.1 is that if there are bars connecting a figure of type A both to a figure type B and a figure type C, then every figure that is of both type B and type C is a figure of type A. That is, this diagram is not only an *inclusion* diagram but also an *intersection* diagram. Specifically, four of the five types of figures in the bottom three tiers of the diagram are intersections of types above them. Here are the four intersections:

*Squares* are exactly those figures that are both rectangles and rhombuses.

*Rectangles* are exactly those figures that are both isosceles trapezoids and parallelograms.

*Rhombuses* are exactly those figures that are both kites and parallelograms.

*Isosceles trapezoids* are exactly those figures that are both cyclic quadrilaterals and trapezoids.

Furthermore, rhombuses are exactly those figures that are kites in two ways, and parallelograms are exactly those figures that are trapezoids in two ways.

Trapezoids occupy a central position in this inclusive hierarchy. Five other special types of figures are trapezoids, so any property of trapezoids becomes a property of all those types. Furthermore, those figures that are both trapezoids and cyclic quadrilaterals are isosceles trapezoids, and if a figure is both a trapezoid and a kite, then it is a parallelogram (but not conversely). So trapezoids have critical relationships with all of the other seven special types of quadrilaterals.

If the exclusive definition of trapezoid is used, then there are still four intersections, so the hierarchy maintains its special character, but one of these intersections has changed from that in the inclusive hierarchy. This is shown in Figure 11.2.

*Squares* are exactly those figures that are both rectangles and rhombuses.

*Rectangles* are exactly those figures that are both cyclic quadrilaterals and parallelograms.

*Rhombuses* are exactly those figures that are both kites and parallelograms.

*Isosceles trapezoids* are exactly those figures that are both cyclic quadrilaterals and trapezoids.

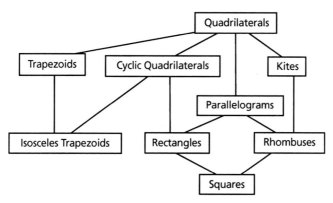

**Figure 11.2** An exclusive quadrilateral hierarchy with eight special types of quadrilaterals.

Now the central position in the hierarchy is occupied by the parallelogram. Parallelograms that are cyclic quadrilaterals are rectangles, and parallelograms that are kites are rhombuses. So properties of parallelograms become paramount in the study of quadrilaterals. When cyclic quadrilaterals and kites are not studied, three of these intersections do not appear. Yet the one remaining intersection is that figures that are both rectangles and rhombuses are squares, and all of these figures are parallelograms. The place of parallelograms in the exclusive hierarchy is why in the majority of today's high school geometry books within the U.S. properties of the parallelogram are given more attention than properties of other quadrilaterals.

The difference in the position of the trapezoid in the hierarchies of Figures 11.1 and 11.2 is striking. In the hierarchy of Figure 11.1, which tends to be preferred by geometers, deducing properties of trapezoids is a way of avoiding work with other special types. In the hierarchy of Figure 11.2, because trapezoids are so isolated, work with trapezoids has little application to other quadrilaterals. A figure cannot be a trapezoid and a kite, because then it would be a parallelogram, and trapezoids cannot be parallelograms.

*Thus the choice of definition for trapezoid influences the amount of attention one should give to the various types of quadrilaterals.*

Figure 11.2 also demonstrates how cyclic quadrilaterals help to tighten up the hierarchy. Without them, isosceles trapezoids are isolated and rectangles are not intersections of two other figures. Yet most geometry texts do not discuss cyclic quadrilaterals. This analysis again suggests that cyclic quadrilaterals deserve more attention, particularly when an exclusive definition of trapezoid is used.

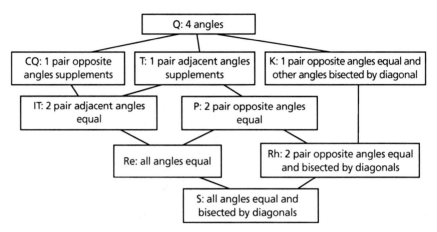

**Figure 11.3** Quadrilaterals in the inclusive hierarchy defined by properties of their angles.

## HIERARCHICAL PROPERTIES OF ANGLES

It is interesting to examine the inclusive hierarchy from the standpoint of properties of these quadrilaterals. We begin by considering properties of angles that can define these types of quadrilaterals. In Figure 11.3 we have replaced the full names of types of quadrilaterals by abbreviations and present defining properties of those quadrilaterals in terms of angles.

The properties shown in Figure 11.3 are not necessarily the only defining properties of quadrilaterals in terms of angles. For example, a defining property of quadrilaterals as polygons is that the sum of the measures of the interior angles is 360°. A defining property of parallelograms is that there are two overlapping pairs of adjacent supplementary angles.

## HIERARCHICAL PROPERTIES OF SIDES

A hierarchy somewhat similar to that of Figure 11.3 appears when considering defining properties of quadrilaterals that involve sides. We show this in Figure 11.4. Figure 11.4 states in the language of distance that there is a circle (with center O) containing all the vertices of the quadrilateral. This idea can be continued down the left side of the hierarchy. Isosceles trapezoids can be defined as those cyclic quadrilaterals for which AB = CD. Rectangles become those cyclic quadrilaterals (or isosceles trapezoids) in which $\overline{AC}$ and $\overline{BD}$ contain O. Squares can be defined as those rectangles in which $\overline{AC} \perp \overline{BD}$.

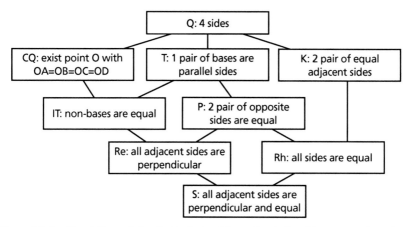

**Figure 11.4**  Quadrilaterals in the inclusive hierarchy defined by properties of their sides.

## HIERARCHICAL PROPERTIES OF DIAGONALS

Each special type of quadrilateral that we have discussed also can be defined in terms of a property of its diagonals, as shown in Figure 11.5.

In Figure 11.5, we note that the diagonals of a trapezoid intersect proportionally. By this we mean that if the quadrilateral is ABCD and the intersection of the diagonals is point E, then $\frac{AE}{CE} = \frac{BE}{DE}$, as shown in Figure 11.6. This proportion is true if and only if $\triangle ADE \sim \triangle CBE$, which in turn is true if and only if m$\angle$ADE = m$\angle$CBE, which itself is true if and only if $\overline{AD} \parallel \overline{BC}$.

The diagonals of a convex quadrilateral separate the quadrilateral into four triangles. The quadrilaterals can also be defined by the shapes and rela-

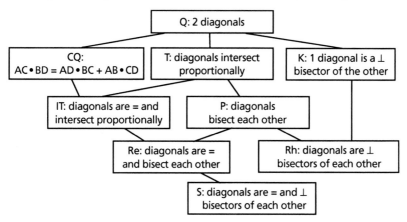

**Figure 11.5**  Quadrilaterals in the inclusive hierarchy defined by properties of their diagonals.

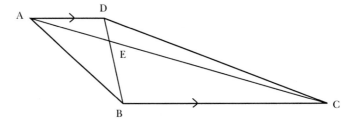

**Figure 11.6** A trapezoid and its diagonals.

tive positions of these triangles. For instance, the triangles ADE and CBE in Figure 11.6 are not only similar, but directly similar. One triangle can be mapped onto the other by the composite of a size change (dilatation) and a rotation: no reflection is needed. If a reflection is needed to map one figure onto a congruent or similar figure, then the figures are oppositely similar. Direct or opposite similarity can be easily determined by imagining walking around the triangles in the orders of their vertices. If the walk is in a clockwise direction for one triangle and in a counterclockwise direction for the other, the triangles are oppositely similar; otherwise, they are directly similar.[1]

In the descriptions in Figure 11.7, we require that under a congruence or similarity correspondence, the intersection of the diagonals corresponds to itself.

If the exclusive definition of trapezoid is employed, the hierarchies of Figures 11.3, 11.4, 11.5, and 11.7 hold for the properties of the figures but the property of a trapezoid is no longer a defining condition for the trapezoid.

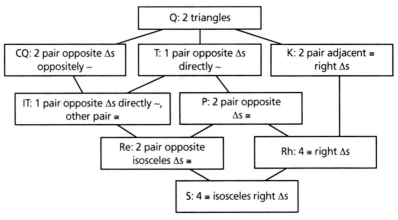

**Figure 11.7** Quadrilaterals in the inclusive hierarchy defined by properties of the triangles formed by their diagonals.

---

1. For the order ADE and the corresponding order CBE in the triangles of Figure 11.6, the walks are both clockwise.

## HIERARCHICAL PROPERTIES OF SYMMETRY

The symmetries of these quadrilaterals are related in a very nice way. Cyclic quadrilaterals and trapezoids, possessing no symmetry, are not involved. Figure 11.8 shows the hierarchy in a form similar to the earlier figures, while Figure 11.9 shows the symmetry using pictures of the identified quadrilaterals.

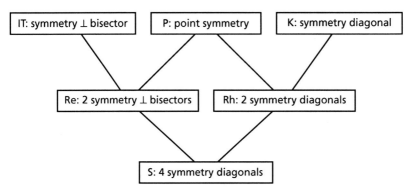

**Figure 11.8** Quadrilaterals in the inclusive hierarchy defined by their symmetries.

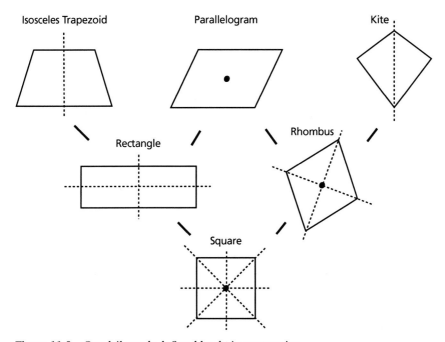

**Figure 11.9** Quadrilaterals defined by their symmetries.

## HIERARCHICAL POSITIONS ON THE COORDINATE PLANE

We did not undertake an exhaustive analysis of the treatment of coordinate geometry in secondary school texts, but it is our impression that in many books the placements of figures on the coordinate plane are assumed rather than deduced. For instance, a book may ask a student to deduce that both pairs of opposite sides of the quadrilateral with vertices $(0,0)$, $(a,0)$, $(a+b,c)$, $(b,c)$ are parallel. In this way, the student has proved that the quadrilateral is a parallelogram. But, even when the book may use these vertices to deduce (algebraically) various properties of parallelograms, rarely does the book demonstrate or ask for a demonstration that, given a parallelogram, the coordinate plane can be located so that the vertices of the parallelogram are of this form. When no discussion regarding the sufficiency of the form takes place, the parallelogram has been implicitly defined as a figure whose vertices can be of this form. Similar implicit definitions are given for many of the other quadrilaterals studied in coordinate geometry sections of geometry texts. Thus our examination of definitions for quadrilaterals includes a study of the positions of quadrilaterals on the coordinate plane.

Figures 11.10, 11.11, 11.12, and 11.14 depict quadrilateral hierarchies on the coordinate plane. In each figure we use the minimal number of variables necessary to determine the quadrilateral and assume restrictions on the variables so that the vertices are distinct.[2] The equations on the segments connecting a type of figure to a second type below it indicate how the lower figure is a special case of the higher one.

In Figure 11.10, the quadrilaterals are placed in what is sometimes called *standard position*, with the picture of the quadrilateral as much as possible in the first quadrant, even though if the values of any of the variables are negative, the quadrilateral would be located in other quadrants. The positions of the trapezoid and its special cases are found in most geometry textbooks that discuss coordinate geometry. However, we have not noticed the positions presented here of the kite and rhombus in the texts we examined (though we did not undertake a complete search); the positions here were shown to us by Doris Schattschneider.

Figure 11.11 shows a hierarchy of those quadrilaterals that are trapezoids under inclusive definitions. The positions emphasize symmetry to the $y$-axis and the rotation symmetry of the parallelogram. In both Figures 11.10 and 11.11, parallelograms are quite naturally special cases of trapezoids. The inclusive definition is so natural that it seems that most geometry books with an exclusive definition for trapezoid (as a figure with exactly one pair of

---

2. It is interesting to ask what the properties of the various types of quadrilaterals become when two of the vertices of the quadrilaterals are allowed to coincide, thus forming a triangle, but we have not undertaken that analysis.

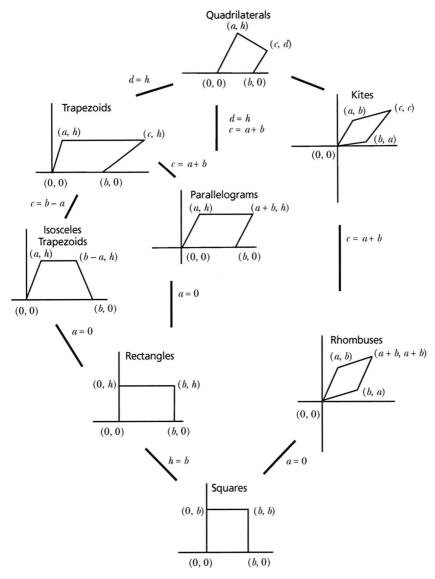

**Figure 11.10** Hierarchy of quadrilaterals in convenient first quadrant positions of the coordinate plane.

parallel sides) ignore the exclusivity when discussing coordinate geometry. The exclusive definition would require the restriction $c \neq a + b$ for the trapezoid in Figure 11.10 and $c \neq d + 2a$ for the trapezoid in Figure 11.11.

Similarly, an exclusive definition for isosceles trapezoid (one that disallows rectangles) would require $a \neq 0$ for the isosceles trapezoid in Figure

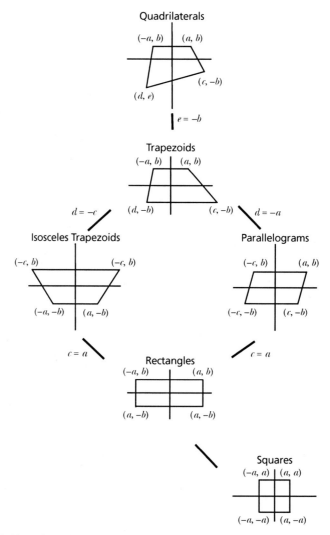

**Figure 11.11** Hierarchy of quadrilaterals in symmetric positions about the axes and origin on the coordinate plane emphasizing horizontal and/or vertical sides.

11.10 and $c \neq a$ for the isosceles trapezoid in Figure 11.11. In general, the characterizations of the various types of quadrilaterals on the coordinate plane provide very strong rationales for inclusive definitions.

Figure 11.12 shows a hierarchy of those quadrilaterals that are kites under inclusive definitions. The positions emphasize symmetry to diagonals. We have also included a position of a parallelogram we have not seen in textbooks to show how the rhombus can be seen as a special case of both the parallelogram and kite. If a definition of kite is offered that does not include

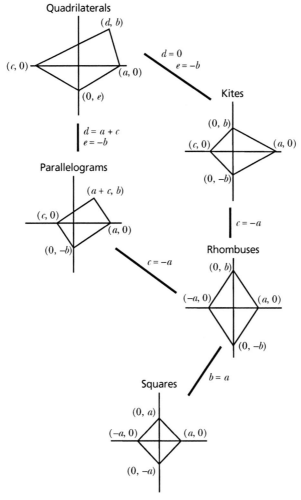

**Figure 11.12**   Hierarchy of quadrilaterals in symmetric positions about the axes and origin in the coordinate plane emphasizing horizontal and/or vertical diagonals.

rhombuses, then the restriction $c \neq a + b$ must be placed on the kite in Figure 11.10 and the restriction $c \neq -a$ must be placed on the kite in Figure 11.12.

In Figures 11.10 and 11.12, the kites can be non-convex or degenerate (forming an isosceles triangle as discussed in Chapter 7 of this monograph). In Figure 11.10, if $c < \frac{a+b}{2}$, and in Figure 11.12, if $ac < 0$, then the kite is not convex. In Figure 11.10, if $c = \frac{a+b}{2}$, and in Figure 11.12, if $ac = 0$, then a degenerate kite appears.

From Figures 11.10 and 11.12, one advantage of studying kites becomes clear. Their properties are easily deduced using coordinates and the properties of rhombuses fall out as a special case. To determine the coordinates for the placement of the kite in both figures, we use the property that one diagonal of a kite is the perpendicular bisector of the other and a symmetry diagonal for the kite. In Figure 11.10 we locate the coordinate system so that the symmetry diagonal is on the line $y = x$ and one vertex is at $(0,0)$. Then the other two vertices of the kite are symmetric to that line and have vertices of the form $(a, b)$ and $(b, a)$. In Figure 11.12, we can locate a coordinate system so that the origin is the intersection of the kite's diagonals and so that the diagonals lie on the coordinate axes. Then the vertices of the kite must lie on the axes (they are on the diagonals), and if one endpoint of the diagonal that is not identified as the symmetry diagonal is $(0, b)$, the other endpoint must be the same distance from the origin on the other side, so must be $(0, -b)$.

When two constraints on a figure lead to the figure below it, as happens from quadrilaterals to kites in Figure 11.12, or from quadrilaterals to parallelograms in the same figure, it is natural to consider the type of figure that is formed when only one constraint is applied.

For instance, when $d = 0$ in the quadrilateral of Figure 11.12, then the vertices of the quadrilateral become $(a, 0)$, $(0, b)$, $(c, 0)$, and $(0, d)$. The diagonals of this quadrilateral are perpendicular, and any quadrilateral with perpendicular diagonals can be placed with these vertices. Altshiller-Court calls such a quadrilateral an orthodiagonal quadrilateral.[3] Orthodiagonal quadrilaterals have some nice rather easily-deduced properties that make this type of quadrilateral appropriate for problems in school textbooks.

> **Theorem OD1:** The sum of the squares of two opposite sides of any orthodiagonal quadrilateral is equal to the sum of the squares of the other two sides.

> **Theorem OD2:** The quadrilateral formed by joining midpoints of consecutive sides of an orthodiagonal quadrilateral is a rectangle.

> **Theorem OD3:** The area of any orthodiagonal quadrilateral is equal to one-half the product of the lengths of its diagonals.

Theorems OD1-OD3 hold for all kites and all rhombuses, justifying the inclusive definitions for these figures. Altshiller-Court also deduces a num-

---

3. Altshiller-Court, *College Geometry*, p. 136.

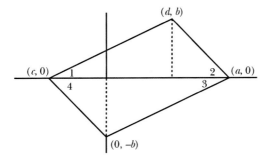

**Figure 11.13**  A quadrilateral with an area-bisecting diagonal.

ber of theorems about those quadrilaterals that are both orthodiagonal and cyclic, but does not give these a special name.[4]

When the constraint $e = -b$ is placed on the quadrilateral of Figure 11.12, then the quadrilateral has vertices $(a,0)$, $(d,b)$, $(c,0)$, and $(0,-b)$ as shown in Figure 11.13. The dotted segments in Figure 11.13, the perpendiculars to a diagonal from opposite vertices, have the same length. Thus the two triangles formed by the sides and this diagonal have the same area. Having an area-bisecting diagonal is a defining characteristic of this type of quadrilateral. This quadrilateral becomes a kite if $d = 0$ or, equivalently, if angles 1 and 2 have the same measure. It becomes a parallelogram if $d = a + c$ or, equivalently, if angles 1 and 3 have the same measure.

Figure 11.14 shows a hierarchy of placements on the coordinate plane for those quadrilaterals that are cyclic quadrilaterals. The origin of the coordinate system is the center of the circle containing the four vertices. Any point on a circle with center at the origin and radius r has coordinates of the form $(r\cos\phi, r\sin\phi)$ from which the coordinates of all vertices of these cyclic quadrilaterals can be determined. The analysis shows again how rectangles can be viewed as special cases of isosceles trapezoids and provides an unusual orientation for both figures. This hierarchy is not found in school geometry texts, for it requires trigonometry as well as mention of cyclic quadrilaterals.

With this hierarchical analysis, we end our study of definitions of quadrilaterals. We encourage reactions from readers that point out errors in our analysis, new insights, analogous analyses of definitions of other mathematical objects, and any other comments that might be of use to those engaged in the study of mathematics curriculum.

---

4.  Ibid., pp. 137–138.

### Cyclic Quadrilaterals

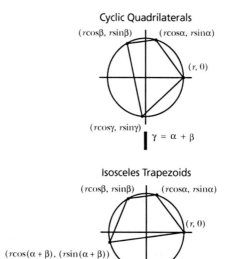

$(r\cos\beta,\ r\sin\beta)$     $(r\cos\alpha,\ r\sin\alpha)$

$(r, 0)$

$(r\cos\gamma,\ r\sin\gamma)$

$\gamma = \alpha + \beta$

### Isosceles Trapezoids

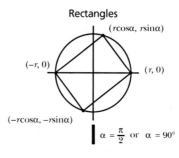

$(r\cos\beta,\ r\sin\beta)$     $(r\cos\alpha,\ r\sin\alpha)$

$(r, 0)$

$(r\cos(\alpha + \beta),\ (r\sin(\alpha + \beta))$

$\beta = \pi$ or $\beta = 180°$

### Rectangles

$(r\cos\alpha,\ r\sin\alpha)$

$(-r, 0)$

$(r, 0)$

$(-r\cos\alpha,\ -r\sin\alpha)$

$\alpha = \dfrac{\pi}{2}$ or $\alpha = 90°$

### Squares

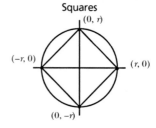

$(0, r)$

$(-r, 0)$

$(r, 0)$

$(0, -r)$

**Figure 11.14**   A hierarchy of cyclic quadrilaterals on the coordinate plane.

# BIBLIOGRAPHY

Altshiller-Court, Nathan. *College Geometry*. New York: Barnes and Noble, 1925, 1952.

Bennett, Albert B. Jr. and L. Ted Nelson. *Mathematics for Elementary Teachers: A Conceptual Approach*. 3rd ed. Dubuque, IA: W. C. Brown, 1992.

Coxeter, H. S. M. *Introduction to Geometry*. New York: Wiley, 1961.

Coxeter, H. S. M. and S. L. Greitzer. *Geometry Revisited*. Washington: Mathematical Association of America, 1967. (Originally published by Random House, 1967.)

Craine, Timothy V. and Rheta N. Rubenstein. "A Quadrilateral Hierarchy to Facilitate Learning in Geometry." *Mathematics Teacher* 86 (January 1993) 30–36.

Dunham, Douglas. "Families of Escher Patterns." In *M.C. Escher's Legacy: A Centennial Celebration: Collection of articles coming from the M.C. Escher Centennial Conference, Rome 1998*, edited by Doris Schattschneider and Michele Emmer, pp. 286–296. New York: Springer-Verlag, 2003a.

———. "Hyperbolic Arc and the Poster Pattern", Math Awareness Month, http://mathforum.org/mam.03/essay1.html (April 2003b)

Edwards, Barbara S. and Michael B. Ward. "Surprises from Mathematics Education Research: Student (Mis)use of Mathematical Definitions." *The American Mathematical Monthly*. Vol. 111, no. 5 (May 2004) 411–424.

Euclid. *Elements*. Edited and translated with introduction and commentary by Sir Thomas Heath. New York: Dover, 1956.

Eves, Howard. *A Survey of Geometry*. Revised Edition. Boston, MA: Allyn and Bacon, 1972.

Forbes, Jack E. and Robert E. Eicholz. *Mathematics for Elementary Teachers*. Reading, MA: Addison-Wesley, 1971.

Gans, David. *Transformations and Geometries*. New York: Appleton-Century-Crofts, 1969.

Greenburg, Marvin J. *Euclidean & Non-Euclidean Geometry: Development and History.* San Francisco: W. H. Freeman & Company, 1974.

Johnson, Roger A. *Advanced Euclidean Geometry.* New York: Dover, 1960.

Keedy, M.L. "What is a Trapezoid?" *Mathematics Teacher* 59 (November 1966) 488–489.

Maraldo, Stephen. "Properties of Quadrilaterals." *The Mathematics Teacher* 73 (January 1980) 38, 39.

Math Forum, definitions and classifications of quadrilaterals, http://mathforum.org/epigone/geom.pre-college/freechermzhor, and http://mathforum.org/epigone/geometry-pre-college/96.

Math Forum, trapezoid definition discussions, http://mathforum.org/library/drmath/view/54901.html.

Miller, Jeff. "Earliest Known Uses of Some of the Words of Mathematics." http://members.aol.com/jeff570/mathword.html

Moise, Edwin E. *Elementary Geometry from an Advanced Standpoint.* Reading, MA: Addison-Wesley, 1963.

National Committee on Mathematical Requirements, *The Reorganization of Mathematics in Secondary Education.* Boston, MA: Houghton Mifflin, 1923, 1927.

Numericana, http://home.att.net/~numbericana/answer/culture.htm#bars.

Pereira-Mendoza, Lionel. "What is a Quadrilateral?" *Mathematics Teacher* 86 (December 1993) 774–777.

Prevost, Fernand J. "Geometry in the Junior High School." *Mathematics Teacher* 78 (September 1985) 411–418.

Quadrilateral, *Merriam Webster's Collegiate Dictionary.* 10th ed. Springfield, MA: Merriam-Webster, 1993.

Rhombus origin, http://www.pballew.net/rhomb.html.

Robinson, Richard. *Definition.* London, ENG: Oxford University Press, 1954.

Schattschneider, Doris. *Visions of Symmetry: Notebooks, Periodic Drawings and Related Work of M.C. Escher.* New Edition. New York: W. H. Freeman & Co., 1990.

Schwartzman, Steven. *The Words of Mathematics.* Washington, DC: Mathematical Association of America, 1994.

Van Hiele, Pierre M. *Structure and Insight: A Theory of Mathematics Education.* Orlando, FL: Academic Press, 1986.

Whiteley, Walter. *Teaching to See like a Mathematician.* 2002. www.math.yorku.ca/Who/Faculty/Whiteley/Teaching_to_see.pdf

Wolfram Research, http://mathworld.wolfram.com

# DEFINITIONS OF QUADRILATERALS

The following pages catalog the definitions of types of quadrilaterals found in the 124 textbooks examined for this monograph. These books were written for three different courses: a high school geometry course or its equivalent; a college course for prospective elementary school teachers; or a college-level geometry course for mathematics majors. The texts are arranged chronologically by date of publication, with the most recent text first, beginning with the high school texts, continuing with the elementary teacher texts, and ending with the texts for the mathematics major.

Each of the 124 texts was examined individually to find and categorize the definitions given to the following: quadrilateral, polygon, parallelogram, trapezoid, rectangle, isosceles trapezoid, kite, rhombus, square, cyclic quadrilateral, trapezium and rhomboid.

The following are the abbreviations given for the quadrilaterals:

| | |
|---|---|
| Q | quadrilateral |
| Re | rectangle |
| S | squares |
| PG | polygon |
| IT | isosceles trapezoid |
| CQ | cyclic quadrilateral |
| P | parallelogram |

*The Classification of Quadrilaterals: A Study of Definition*, pages 85–90

| | |
|---|---|
| K | kite |
| Trm | trapezium |
| T | trapezoid |
| Rh | rhombus |
| Rhd | rhomboid |

Multiple editions of texts that share a lead author and define each of the quadrilaterals in the same way are listed on the same line. The "Year" given for a text is the latest year listed on the copyright page. Dates for all but the latest editions examined can be found under the column heading "other editions." Multiple editions of texts that share a lead author and define one or more of the quadrilaterals differently are listed directly below the latest edition so as to provide easy comparison.

With the exception of "polygon," only definitions in the high school texts are counted and discussed in the monograph. Multiple editions of a single text that defined the quadrilaterals in the same way are counted as one. Editions with different defining conditions for one or more of the quadrilaterals are counted separately because it is clear the authors re-examined the definitions they used.

A complete list of the definitions found in the 124 texts is given in Appendix A. The numbers listed for these definitions correspond to the numbers found in the columns on the following pages as well as those in the monograph. A detailed bibliography of the 124 texts can be found in Appendix B.

## 1. QUADRILATERAL

Q(1)   a four-sided polygon
Q(2)   the union of four line segments that join four coplanar points, no three of which are collinear, each segment intersecting exactly two others, one at each endpoint
Q(3)   a closed, four-sided plane figure
Q(4)   a portion of a plane bounded by four sects
Q(5)   a closed figure with four straight line segments
Q(6)   a simple polygon with four sides
Q(7)[**]   four points, A, B, C and D are coplanar and no three of them are collinear

## 2. POLYGON

PG(1)  the union of three or more coplanar segments such that each segment intersects exactly two other segments, one at each endpoint, and no two intersecting segments are collinear

PG(2)  a closed plane figure consisting of line segments called sides

PG(3)  a portion of a plane bounded by straight lines

PG(4)  a closed broken line

PG(5)  a portion of a plane bounded by line segments of which consecutive pairs do not lie on the same line

PG(6)  a simple closed polygonal curve

PG(7)  the union of three or more coplanar segments which intersect only at their endpoints and each endpoint is shared exactly by two others

PG(8)  a closed broken line in a plane

PG(9)  any rectilinear figure, regardless of the number of sides

PG(10)  a simple closed curve that is the union of segments which intersect only at their endpoint, each endpoint is shared by exactly two segments, and such that the endpoints of segments sharing an endpoint are not on the same line

PG(11)  a closed geometric figure in a plane formed by connecting line segments endpoint to endpoint with each segment intersecting exactly two others

PG(12)  a figure formed by joining (or fitting) segments at their endpoints in such a way that each segment meets two other segments, one at each endpoint

PG(13)  a closed figure bounded by three or more line segments that intersect exactly two other segments, one at each endpoint

PG(14)  a figure made up of coplanar segments called sides, each intersect exactly two other sides, one at each endpoint, and no two sides with a common endpoint are collinear

PG(15)  a plane figure bounded by finite straight lines

PG(16)  the union of $\overline{P_1P_2}, \ldots, \overline{P_{n-1}P_n}$ and $\overline{P_nP_1}$, where $n > 2$, $P_1, P_2, \ldots, P_n$ are $n$ different points, and no three consecutive points are collinear

PG(17)  a closed figure in a plane which is the union of line segments

PG(18)  a closed figure in a plane which is the union of line segments such that the segments intersect only at their endpoints and no segments sharing a common endpoint are collinear

PG(19)  an $n$-sided figure consisting of $n$ segments $\overline{P_1P_2}, \overline{P_2P_3}, \overline{P_3P_4}, \ldots, \overline{P_{n-1}P_n}, \overline{P_nP_1}$ which intersect only at their endpoints and enclose a single region

PG(20)  a closed two-dimensional shape formed by three or more line segments

PG(21)     a closed plane figure whose sides are segments that intersect only at their endpoints and no two sides with a common endpoint are collinear

PG(22)*    a simple closed curve formed by the union of a finite number of line segments, no two of which in succession are collinear

PG(23)*    a simple, closed polygonal curve such that no two segments with a common endpoint are collinear

PG(24)*    a finite set of segments with the following properties:
- the intersection of any pair of segments is either the null set or a common endpoint of the pair of segments
- if the intersection of a pair of segments is a common endpoint, then the segments do not lie on the same line
- each endpoint of the segment is common to exactly one other segment

PG(25)*    a simple closed polygonal curve which is the union of line segments in the same plane

PG(26)*    a closed curve in a plane which is the union of three or more line segments placed end to end in a sequence

PG(27)*    a simple closed curve which has a boundary made up entirely of line segments

PG(28)*    a simple closed curve made up entirely of straight line segments

PG(29)**   a simple closed curve that is the union of line segments

## 3. PARALLELOGRAM

P(1)     a quadrilateral with two pairs of parallel sides
P(2)     a trapezoid with two pairs of parallel sides

## 4. TRAPEZOID

T(1)     a quadrilateral with exactly one pair of parallel sides
T(2)     a quadrilateral that has at least one pair of parallel sides

## 5. RECTANGLE

Re(1)     a parallelogram with four right angles
Re(2)     a parallelogram in which at least one angle is a right angle
Re(3)     an equiangular parallelogram
Re(4)     a quadrilateral which has four right angles
Re(5)     an equiangular quadrilateral

Re(6)   a rhomboid with a right angle

Re(7)   a parallelogram whose adjacent sides are unequal and whose angles are right angles

Re(8)   a quadrilateral with three right angles

## 6. ISOSCELES TRAPEZOID

IT(1)   a trapezoid in which the nonparallel sides (legs) are congruent

IT(2)   a trapezoid which has a pair of base angles equal in measure

IT(3)   a trapezoid which is symmetric about a line through the midpoints of its bases

IT(4)   a trapezoid in which at least one pair of opposite sides are congruent

## 7. KITE

K(1)    a quadrilateral which has two distinct pairs of adjacent sides of the same length

K(2)    a quadrilateral in which exactly one diagonal is a perpendicular bisector of the other

K(3)    a quadrilateral in which one diagonal is a perpendicular bisector of the other

K(4)    a quadrilateral with two pairs of adjacent sides congruent and no opposite sides congruent

K(5)    a convex quadrilateral with two distinct pairs of adjacent sides of the same length

K(6)    a 4-side with an axis of symmetry

K(7)*   a quadrilateral with two pairs of congruent sides and no pair of parallel sides

## 8. RHOMBUS

Rh(I1)   a parallelogram with four equal sides

Rh(I2)   a parallelogram in which at least two consecutive sides are congruent

Rh(I3)   a quadrilateral in which all four sides are equal in length/congruent

Rh(E1)   a rhomboid having all sides equal

Rh(E2)   a parallelogram whose angles are oblique and sides are equal

Rh(E3)   a rhomboid having two adjacent sides equal

Rh(E4)   a parallelogram with oblique angles and with two adjacent sides equal

## 9. SQUARE

S(1)   a rectangle with four congruent sides
S(2)   a rectangle with a pair of consecutive congruent sides
S(3)   a parallelogram that is both a rectangle and a rhombus
S(4)   a parallelogram with four equal sides and four right angles
S(5)   a parallelogram with one right angle and two adjacent sides congruent
S(6)   a quadrilateral that has four equal sides and four right angles
S(7)   a rhombus with four equal angles
S(8)   a rhombus with a right angle
S(9)**  a quadrilateral that is both a rhombus and a rectangle

## 10. CYCLIC QUADRILATERAL

CQ(1)   a quadrilateral whose four vertices lie on a circle
CQ(2)   a quadrilateral inscribed in a circle with opposite angles supplementary

---

\*   Indicates the definition is found only in a mathematics for elementary school teachers text.
\*\*   Indicates the definition is found only in a college level text.

# APPENDIX B

# INDIVIDUAL GEOMETRY TEXTS AND THE DEFINITIONS FOUND IN THEM

*The Classification of Quadrilaterals: A Study of Definition*, pages 91–98
Copyright © 2008 by Information Age Publishing
All rights of reproduction in any form reserved.

| Year | Other editions | Author | Title | Publisher | Q | PG | P | Re | Rh | S | T | IT | K | CQ | Trm | Rhd |
|---|---|---|---|---|---|---|---|---|---|---|---|---|---|---|---|---|
| **High School Texts** | | | | | | | | | | | | | | | | |
| 2004 | 1998 | Bass et al. | *Geometry* | Prentice Hall | 1 | 3H | 1 | 1 | 1 | 4 | E | 1 | 3 | 0 | | |
| 2003 | | Coxford et al. | *Core-Plus Mathematics, Course 1 & 3* | Everyday Learning | 1 | 10 | 1 | 2 | 2 | 3 | E | 0 | 1 | 0 | | |
| | 1998 | Coxford et al. | *Core-Plus Mathematics, Course 3* | Everyday Learning | 1 | 10 | 1 | 2 | 1 | 6 | 0 | 0 | 0 | 0 | | |
| 2003 | 1997 | Serra | *Discovering Geometry: An Investigative Approach* | Key Curriculum | 1 | 3A | 1 | 3 | 1 | 7,1 | E | 1 | 1 | 1 | | |
| | 1993 | Serra | *Discovering Geometry: An Inductive Approach* | Key Curriculum | 1 | 3A | 1 | 3 | 1 | 7,1 | E | 1 | 0 | 0 | | |
| 2002 | 1991 | Coxford and Usiskin et al. | *UCSMP Geometry* | Scott, Foresman | 1 | 5A | 1 | 4 | 3 | 5 | 1 | 2 | 1 | 0 | | |
| 2000 | | Education Development Center | *Connected Geometry* | Everyday Learning | 0 | 0 | 1 | 0 | 1 | 0 | E | 1 | 1 | 0 | | |
| 1990 | 1978 | Jurgensen et al. | *Geometry* | Houghton Mifflin | 1 | 6C | 1 | 4 | 3 | 5 | E | 1 | 0 | 0 | | |
| | 1973 | Jurgensen et al. | *Modern Basic Geometry* | Houghton Mifflin | 1 | 3D | 1 | 1 | 3 | 1 | E | 1 | 0 | 0 | | |
| | 1972 | Jurgensen et al. | *Modern School Mathematics Geometry* | Houghton Mifflin | 1 | 1A | 1 | 1 | 1 | 2 | E | 1 | 0 | 0 | | |
| | 1963 | Jurgensen et al. | *Modern Geometry Structure and Method* | Houghton Mifflin | 1 | 5A | 1 | 2 | 2 | 2 | E | 1 | 0 | 0 | | |
| 1999 | | CORD | *CORD Geometry* | Globe Fearon | 1 | 6B | 1 | 1 | 1 | 4 | E | 1 | 0 | 0 | | |
| 1998 | | Aichele et al. | *Geometry: Explorations and Applications* | McDougal Littell | 1 | 3B | 1 | 3 | 1 | 4 | E | 1 | 4 | 0 | | |
| 1997 | | Fendel et al. | *IMP* | Key Curriculum | 1 | 12 | 0 | 3 | 1 | 4 | E | 0 | 0 | 0 | | |

| Year | Author | Title | Publisher | | | | | | | | | | |
|---|---|---|---|---|---|---|---|---|---|---|---|---|---|
| 1995 | Burrill et al. | *Merrill Geometry: Applications and Connections* | Glencoe | 1 | 6A | 1 | 4 | 1 | 3 | E | 1 | 0 | 0 |
| 1992 | Elander | *Geometry for Decision Making* | South-Western Publishing | 3 | 3C | 1 | 2 | 2 | 1 | E | 1 | 0 | 0 |
| 1991 | Rhoad et al. | *Geometry for Enjoyment and Challenge* | McDougal Littell | 1 | 11 | 1 | 2 | 1 | 3 | E | 1 | 1 | 3 |
| 1991 | Nichols et al. | *Geometry* | Holt, Rinehart and Winston | 1 | 5B | 1 | 1 | 2 | 1 | E | 1 | 3 | 0 |
| 1974 | Nichols et al. | *Holt Geometry* | Holt, Rinehart and Winston | 1 | 5A | 1 | 2 | 2 | 2 | 1 | 3 | 0 | 0 |
| 1990 | Hirsch et al. | *Geometry* | Scott, Foresman | 1 | 5B | 1 | 1 | 1 | 1 | E | 1 | 2 | 0 |
| 1990 | Kalin and Corbitt | *Geometry* | Prentice Hall | 1 | 6B | 1 | 2 | 2 | 2 | E | 1 | 0 | 0 |
| 1987 | Travers et al. | *Geometry* | Laidlaw | 2 | 3J | 1 | 1 | 1 | 1 | E | 0 | 0 | 0 |
| 1985 | Rising et al. | *Unified Mathematics Book 2* | Houghton Mifflin | 1 | 1B | 1 | 4 | 3 | 5 | E | 1 | 0 | 0 |
| 1985 | Bumby and Klutch | *Mathematics: A Topical Approach Course 1 & 2* | Merrill | 1 | 9A | 1 | 1 | 1 | 3 | E | 1 | 0 | 0 |
| 1984 | Foster et al. | *Merrill Geometry* | Charles E. Merrill | 1 | 6A | 1 | 1 | 3 | 1 | E | 1 | 0 | 0 |
| 1984 | Pilger | *Geometry for Christian Schools* | Bob Jones University | 1 | 9B | 1 | 1 | 1 | 1,7 | E | 0 | 0 | 0 |
| 1983 | Lang and Murrow | *Geometry: A High School Course* | Springer-Verlag | 1 | 1E | 1 | 1 | 1 | 0 | 1 | 1 | 0 | 0 |
| 1982 | Chakerian et al. | *Geometry: A Guided Inquiry* | Sunburst Communications | 1 | 6E | 1 | 1 | 1 | 0 | 1 | 0 | 0 | 0 |
| 1981 | Keenan and Dressler | *Integrated Mathematics Course II* | AMSCO | 1 | 3I | 1 | 2 | 2 | 2 | E | 1 | 0 | 0 |
| 1973 | Dressler | *Geometry* | AMSCO | 1 | 3C | 1 | 2 | 2 | 2 | E | 1 | 0 | 0 |
| 1978 | Wells et al. | *Using Geometry* | Laidlaw Brothers | 2 | 6D | 1 | 2 | 1 | 1,8 | E | 1 | 1 | 0 |
| 1977 | Posamentier | *Geometry: Its Elements and Structure* | McGraw-Hill | 1 | 1A | 1 | 4 | 2 | 6 | E | 1 | 0 | 1,3 |

| Year | Other editions | Author | Title | Publisher | Q | PG | P | Re | Rh | S | T | IT | K | CQ | Trm | Rhd |
|---|---|---|---|---|---|---|---|---|---|---|---|---|---|---|---|---|
| 1975 | | Coxford and Usiskin | *Geometry: A Transformation Approach* | Laidlaw Brothers | 1 | 1H | 1 | 8 | 3 | 9 | 1 | 1 | 1 | 0 | | |
| 1974 | 1968 | Wilcox | *Geometry: A Modern Approach* | Addison-Wesley | 2 | 1A | 1 | 1 | 1 | 1 | E | 0 | 0 | 3 | | |
| 1974 | | Jacobs | *Geometry* | W. H. Freeman | 3 | 1A | 1 | 5 | 3 | 5 | E | 1 | 1 | 1 | | |
| 1972 | 1969 | Ulrich and Payne | *Geometry* | Harcourt, Brace, Jovanovich | 6 | 9C | 1 | 2 | 2 | 2 | E | 1 | 0 | 0 | | |
| 1972 | 1965 | Ladd and Kelly | *Elements of Geometry* | Scott, Foresman | 1 | 1D | 1 | 5 | 3 | 3 | E | 1 | 0 | 3 | | |
| 1972 | | Jacobs and Meyer | *Discovering Geometry* | Harcourt, Brace, Jovanovich | 1 | 5C | 1 | 1 | 1 | 1 | E | 1 | 0 | 0 | | |
| 1971 | 1966 | Rosskopf et al. | *Geometry* | Silver Burdett | 1 | 1A | 1 | 2 | 2 | 3 | E | 1 | 0 | 3 | | |
| 1971 | | Pearson and Smart | *Geometry* | Ginn | 2 | 1A | 1 | 2 | 2 | 3 | E | 1 | 0 | 0 | | |
| 1971 | | Moise and Downs | *Geometry* | Addison-Wesley | 2 | 1A | 1 | 1 | 1 | 1 | E | 1 | 2 | 3 | | |
| 1971 | 1964 | Moise and Downs | *Geometry* | Addison-Wesley | 2 | 1A | 1 | 1 | 1 | 1 | 1 | 4 | 2 | 2 | | |
| 1970 | | Carico et al. | *Geometry* | Macmillan | 1 | 1A | 1 | 2 | 2 | 1 | E | 1 | 0 | 0 | | |
| 1969 | 1966 | Anderson et al. | *School Mathematics Geometry* | Houghton Mifflin | 2 | 1A | 1 | 1 | 1 | 1 | E | 1 | 0 | 0 | | |
| 1967 | | Keedy et al. | *Exploring Geometry* | Holt Rinehart and Winston | 2 | 1F | 2 | 2 | 1 | 1 | 1 | 0 | 0 | 0 | | |
| 1965 | | Fischer and Hayden | *Geometry* | Allyn and Bacon | 2 | 0 | 1 | 0 | 3 | 0 | E | 1 | 1 | 0 | | |
| 1964 | | Lewis | *Geometry: A Contemporary Course* | D. Van Nostrand | 1 | 1G | 1 | 2 | 2 | 2 | E | 1 | 0 | 0 | | |
| 1963 | | Morgan and Zartman | *Geometry: Plane, Solid, Contintue* | Houghton Mifflin | 1 | 4A | 1 | 2 | 1 | 1 | E | 1 | 0 | 0 | | |
| 1963 | | Kenner et al. | *Concepts of Mathematics Book 2* | American Book | 3 | 0 | 1 | 1 | 1 | 6 | E | 1 | 0 | 0 | | |
| 1961 | | Weeks and Adkins | *A Course in Geometry: Plane and Solid* | Ginn | 1 | 1C | 1 | 2 | 2 | 2 | E | 1 | 0 | 0 | | |

| Year | (orig.) | Author | Title | Publisher | (1) | (2) | (3) | (4) | (5) | (6) | (7) | (8) | (9) | (10) | (11) | (12) | (13) |
|---|---|---|---|---|---|---|---|---|---|---|---|---|---|---|---|---|---|
| 1961 | | School Mathematics Study Group | *Geometry* | Yale University Press | 2 | 5A | 1 | 1 | 1 | 1 | 1 | E | 0 | 5 | 2 | | |
| 1959 | 1941 | Birkhoff and Beatley | *Basic Geometry* | Scott, Foresman | 1 | 4B | 1 | 3 | 1 | 1 | 4 | E | 0 | 0 | 0 | | |
| 1959 | | Hart et al. | *Plane Geometry and Supplements* | D. C. Heath | 1 | 4A | 1 | 2 | 7 | 2 | 2 | E | 1 | 0 | 0 | | |
| 1958 | | Herberg and Orleans | *A New Geometry for Secondary Schools* | D. C. Heath | 5 | 13 | 1 | 1 | 1 | 1 | 4 | 1 | 1 | 0 | 0 | | |
| 1958 | | Welchons et al. | *Plane Geometry* | Ginn | 1 | 4A | 1 | 2 | 2 | 2 | 2 | E | 1 | 0 | 0 | | |
| 1957 | | Schacht and McLennan | *Plane Geometry* | Henry Holt | 1 | 4A | 1 | 2 | 2 | 2 | 2 | E | 1 | 0 | 0 | | |
| 1951 | 1936 | Avery | *Plane Geometry* | Allyn and Bacon | 3 | 2A | 1 | 1 | 1 | 1 | 1 | E | 1 | 0 | 0 | | |
| 1950 | | Welkowitz et al. | *Geometry: Meaning and Mastery* | John C. Winston | 1 | 2B | 1 | 5 | 3 | 3 | 3 | E | 1 | 0 | 0 | | |
| 1950 | | Reichgott and Spiller | *Today's Geometry* | Prentice-Hall | 4 | 3F | 1 | 1 | 1 | 1 | 1 | E | 1 | 0 | 0 | | |
| 1948 | | Smith and Marino | *Plane Geometry* | Charles E. Merrill | 1 | 7B | 1 | 1 | 1 | 1 | 1 | E | 1 | 0 | 3 | | |
| 1948 | | Sigley and Stratton | *Plane Geometry* | Dryden | 1 | 2B | 1 | 3 | 5 | 1 | 1 | E | 1 | 0 | 0 | | |
| 1946 | | Keniston and Tully | *Plane Geometry* | Ginn | 1 | 4A | 1 | 4 | 3 | 6 | 6 | E | 1 | 0 | 0 | | |
| 1941 | | Seymour and Smith | *Plane Geometry* | Macmillan | 1 | 4A | 1 | 2 | 2 | 2 | 2 | E | 1 | 0 | 0 | | |
| 1924 | | Schultz and Sevenoak | *Plane Geometry* | Macmillan | 1 | 4 | 1 | 1 | 1 | 1 | 1 | 1 | 1 | 2 | 2 | | |
| 1924 | | Palmer et al. | *Plane Geometry* | Scott, Foresman | 3 | 2B | 1 | 1 | 4 | 4 | 1 | E | 1 | 0 | 0 | 0 | 1 |
| 1924 | 1915 | Palmer and Taylor | *Plane Geometry* | Scott, Foresman | 3 | 3E | 1 | 1 | 4 | 4 | 1 | E | 1 | 0 | 0 | 1 | 1 |
| 1916 | | Durell and Arnold | *Plane Geometry* | Charles E. Merrill | 1 | 4A | 1 | 1 | 1 | 1 | 1 | E | 1 | 0 | 0 | 1 | |
| 1912 | | Shutts | *Plane and Solid Geometry—Suggestive Method* | Atkinson, Mentzer | 4,3 | 0 | 1 | 1 | 4 | 4 | 1 | E | 1 | 0 | 3 | 1 | |
| 1911 | | Hart and Feldman | *Plane Geometry* | American Book | 1 | 2B | 1 | 2 | 6 | 2 | 2 | E | 1 | 0 | 0 | 1 | 3 |
| 1908 | | Lyman | *Plane Geometry* | American Book | 1 | 2A | 1 | 2 | 2 | 2 | 0 | E | 1 | 0 | 0 | 1 | |
| 1904 | | Quinn | *A Socratic Study of Plane Geometry* | C. W. Bardeen | 5 | 2A | 1 | 6, 2 | 5 | 5 | 1 | E | 1 | 0 | 6 | 1 | 4 |

| Year | Other editions | Author | Title | Publisher | Q | PG | P | Re | Rh | S | T | IT | K | CQ | Trm | Rhd |
|------|------|--------|-------|-----------|---|----|---|----|----|---|---|----|---|----|-----|-----|
| 1903 | | Sanders | *Elements of Plane and Solid Geometry* | American Book | 1 | 7A | 1 | 1 | 4 | 1 | E | 0 | 0 | 3 | 1 | 1 |
| 1902 | | Hopkins | *Inductive Plane Geometry* | D. C. Heath | 3 | 0 | 1 | 1 | 1 | 3 | E | 1 | 0 | 0 | 1 | 2 |
| 1901 | | Holgate | *Elementary Geometry: Plane and Solid* | Macmillan | 1 | 8B | 1 | 2 | 2 | 6 | E | 1 | 0 | 3,1 | 1 | |
| 1899 | | Milne | *Plane and Solid Geometry* | American Book | 4 | 7A | 1 | 1 | 4 | 1 | E | 1 | 0 | 0 | 1 | 1 |
| 1897 | | Smith | *School Geometry* | Scott, Foresman | 1 | 2A | 1 | 7 | 5 | 4 | E | 1 | 0 | 0 | 1 | 2 |
| 1896 | | Phillips and Fisher | *Plane Geometry* | American Book | 1 | 2A | 1 | 1 | 3 | 1 | E | 0 | 0 | 0 | 1 | |
| 1896 | | Hobbs | *The Elements of Plane Geometry* | A. Lovell | 4 | 7A | 1 | 1 | 5 | 1 | E | 1 | 0 | 0 | 1 | 2 |
| 1896 | | Pettee | *Plane Geometry* | Silver, Burdett | 1 | 2A | 1 | 2 | 4 | 1 | E | 1 | 0 | 1 | 1 | 3 |
| 1895 | | Hornbrook | *Concrete Geometry for Beginners* | American Book | 0 | 0 | 1 | 1 | 0 | 1 | E | 1 | 0 | 1 | 1 | 6 |
| 1895 | | Macnie | *Elements of Plane Geometry* | American Book | 1 | 7A | 1 | 2 | 5 | 1 | E | 0 | 0 | 0 | 1 | 0 |
| 1894 | | Dupuis | *Elementary Synthetic Geometry* | Macmillan | 3 | 8C | 1 | 2 | 2 | 2 | E | 0 | 0 | 0 | 1 | |
| 1893 | | Smith | *Introductory Modern Geometry of Point, Ray, and Circle* | Macmillan | 1 | 2A | 1 | 3 | 1 | 3 | E | 0 | 0 | 3 | | |
| 1874 | | Phillips | *Elements of Geometry and the First Principles of Modern Geometry* | J. W. Schermerhorn | 1 | 7A | 1 | 2 | 4 | 1 | E | 0 | 0 | 1 | 1 | 1 |
| 1849 | | Perkins | *Elements of Geometry with Practical Applications* | H. H. Hawley | 1 | 8A | 1 | 1 | 5 | 1 | E | 0 | 0 | 0 | 1 | 0 |
| 1833 | | Young | *Elements of Geometry with Notes* | Carey, Lea and Blanchard | 1 | 8A | 1 | 6 | 6 | 8 | 0 | 0 | 0 | 1 | 2 | 5 |

Hopkins (1902) is the only text in this list to define oblong (as a rectangle whose adjacent sides are unequal).

*Additional editions listed that gave the same definitions          Total HST: 86

## Elementary School Teacher Math Texts

| Year | Other editions | Author | Title | Publisher | Q | PG | P | Re | Rh | S | T | IT | K | CQ | Trm | Rhd |
|---|---|---|---|---|---|---|---|---|---|---|---|---|---|---|---|---|
| 1995 | | Wheeler and Wheeler | Modern Mathematics | Brooks/Cole | 1 | 9A | 1 | 1 | 1 | 1 | E | 0 | 1 | 0 | | |
| 1992 | | Bennett and Nelson | Mathematics for Elementary Teachers: A Conceptual Approach | Wm. C. Brown | 1 | 21 | 1 | 4 | 8 | 5 | 1 | 0 | 0 | 0 | | |
| | 1979 | Bennett and Nelson | Mathematics: An Informal Approach | Allyn and Bacon | 1 | 14 | 1 | 1 | 1 | 5 | E | 0 | 0 | 0 | | |
| 1988 | | Musser and Burger | Mathematics for Elementary Teachers: A Contemporary Approach | Macmillan | 3 | 14 | 1 | 4 | 3 | 5 | E | 1 | 1 | 0 | | |
| 1987 | | Krause | Mathematics for Elementary Teachers: A Balanced Approach | D. C. Heath | 1 | 9A | 1 | 1 | 3 | 5,8 | E | 1 | 7 | 0 | | |
| 1982 | | Gerber | Mathematics for Elementary School Teachers | Saunders College Publishing | 1 | 18 | 1 | 1 | 1 | 1 | E | 0 | 1 | 0 | | |
| 1981 | | Billstein et al. | A Problem Solving Approach to: Mathematics for Elementary School Teachers | Benjamin/Cummings | 1 | 15 | 1 | 2 | 1 | 1 | E | 0 | 0 | 0 | | |
| 1977 | 1972 | Weiss | Elementary College Mathematics | Prindle, Weber and Schmidt | 1 | 3F | 1 | 2 | 1 | 8,1 | 1 | 0 | 0 | 0 | | |
| 1975 | | Graham | Modern Elementary Mathematics | Harcourt Brace Javanovich | 1 | 14 | 1 | 1 | 1 | 1 | E | 0 | 0 | 0 | | |
| 1974 | | Devine and Kaufmann | Mathematics for Elementary Education | John Wiley and Sons | 1 | 15 | 1 | 2 | 2 | 2 | E | 1 | 0 | 0 | | |
| 1971 | | Forbes and Eicholz | Mathematics for Elementary Teachers | Addison-Wesley | 1 | 1A | 1 | 2 | 2 | 2 | 1 | 0 | 0 | 0 | | |

| Year | Other editions | Author | Title | Publisher | Q | PG | P | Re | Rh | S | T | IT | K | CQ | Trm | Rhd |
|---|---|---|---|---|---|---|---|---|---|---|---|---|---|---|---|---|
| 1970 | | Brumfiel and Vance | *Algebra and Geometry for Teachers* | Addison-Wesley | 1 | 0 | 1 | 2 | 2 | 0 | E | 0 | 0 | 0 | | |
| | 1962 | Brumfiel et al. | *Fundamental Concepts of Elementary Mathematics* | Addison-Wesley | 1 | 19 | 1 | 1 | 1 | 1 | E | 0 | 0 | 0 | | |
| 1969 | | Ohmer | *Elementary Geometry for Teachers* | Addison-Wesley | 1 | 17 | 1 | 2 | 1 | 2 | 0 | 0 | 0 | 0 | | |
| 1966 | | Byrne | *Modern Elementary Mathematics* | McGraw-Hill | 1 | 16 | 1 | 1 | 0 | 1 | E | 0 | 0 | 0 | | |

Total ESTMT: 15

**College Level Texts**

| Year | Other editions | Author | Title | Publisher | Q | PG | P | Re | Rh | S | T | IT | K | CQ | Trm | Rhd |
|---|---|---|---|---|---|---|---|---|---|---|---|---|---|---|---|---|
| 1986 | | Ryan | *Euclidean and Non-Euclidean Geometry* | Cambridge University Press | 0 | 0 | 0 | 5 | 3 | 7 | 0 | 0 | 0 | 0 | | |
| 1973 | | Allen and Guyer | *Basic Concepts in Geometry: An Introduction to Proof* | Dickenson | 7 | 1F | 1 | 2 | 2 | 6 | E | 1 | 0 | 0 | | |
| 1973 | | Greenberg | *Euclidean and Non-Euclidean Geometries: Development and History* | W. H. Freeman | 2 | 0 | 1 | 4 | 9 | 0 | 0 | 0 | 0 | 0 | 0 | |
| 1971 | | Ringenberg and Presser | *Geometry* | Wiley and Sons | 2 | 1A | 1 | 2 | 2 | 2 | 1 | 0 | 0 | 0 | 1 | |
| | 1968 | Ringenberg | *College Geometry* | John Wiley and Sons | 1 | 0 | 1 | 1 | 1 | 1 | E | 0 | 0 | 0 | | |
| 1972 | | Bouwsma | *Geometry for Teachers* | Macmillan | 1 | 20 | 1 | 4 | 0 | 1 | 0 | 0 | 0 | 0 | | |
| 1969 | | Stubblefield | *An Intuitive Approach to Elementary Geometry* | Brooks/Cole | 2 | 0 | 1 | 5 | 3 | 10 | 1 | 0 | 0 | 0 | | |
| 1963 | | Moise | *Elementary Geometry from an Advanced Standpoint* | Addison-Wesley | 2 | 0 | 1 | 4 | 2 | 0 | 1 | 0 | 0 | 1,3 | | |

Total CLT: 8

# BIBLIOGRAPHY OF TEXTS EXAMINED

Aichele, Douglas B., Patrick Hopfensperger, Miriam A. Leiva, Marguerite M. Mason, Stuart J. Murphy, Micki J. Schell and Matthias C. Vheru. *Geometry: Explorations and Applications.* Evanston, IL: McDougal Littell, 1998.

Allen, Frank B. and Betty Stine Guyer. *Basic Concepts in Geometry: An Introduction to Proof.* Encino, CA: Dickenson Publishing Co., 1973.

Anderson, Richard D., Jack W. Garon and Joseph G. Gremillion. *School Mathematics Geometry.* Boston, MA: Hougton Mifflin Co., 1966, 1969.

Avery, Royal A. *Plane Geometry.* Boston, MA: Allyn and Bacon, 1936, 1951.

Bass, Laurie E., Randall I. Charles, Art Johnson and Dan Kennedy. *Geometry.* Needham, MA: Prentice Hall, 1998, 2004.

Bennett, Albert B. and Leonard T. Nelson. *Mathematics For Elementary Teachers, A Conceptual Approach.* 3rd ed. Dubuque, IA: Wm. C. Brown Publishers, 1992.

_____. *Mathematics: An Informal Approach.* Boston, MA: Allyn and Bacon, 1979.

Billstein, Rick, Shlomo Libeskind and Johnny W. Lott. *A Problem Solving Approach to Mathematics for Elementary School Teachers.* Menlo Park, CA: The Benjamin/Cummings Publishing Co., 1981.

Birkhoff, George David and Ralph Beatley. *Basic Geometry.* New York: Scott, Foresman and Co., 1941, 1959.

Bouwsma, Ward D. *Geometry for Teachers.* New York: The Macmillan Co., 1972.

Brumfiel, Charles F. and Irvin E. Vance. *Algebra and Geometry for Teachers.* Reading, MA: Addison-Wesley Publishing Co., 1970.

---

Brumfiel, Charles F., Robert E. Eicholz and Merrill E. Shanks. *Fundamental Concepts of Elementary Mathematics.* Reading, MA: Addison-Wesley Publishing Co., 1962.

Bumby, Douglas and Richard Klutch. *Mathematics: A Topical Approach, Course 2.* Columbus, OH: Charles E. Merrill Publishing Co., 1978, 1985.

Burrill, Gail F., Jerry J. Cummins, Timothy D. Kanold and Lee E. Yunker. *Merrill: Geometry: Applications and Connections.* New York: Glencoe, 1995.

Byrne, J. Richard. *Modern Elementary Mathematics.* New York: McGraw-Hill Book Co., 1966.

Carico, Charles C., Herman Hyatt, Irving Drooyan and James Hardesty. *Geometry.* New York: Macmillan Co., 1970.

Center for Occupational Research and Development. *CORD Geometry.* Upper Saddle River, NJ: Globe Fearon, 1999.

Chakerian, G.D., Calvin D. Crabill, Sherman K. Stein. *Geometry: A Guided Inquiry.* Boston, MA: Houghton Mifflin Co., 1972, 1982.

Coxford, Arthur F., James T. Fey, Christian R. Hirsch, Harold L. Schoen, Gail Burrill, Eric W. Hart and Ann E. Watkins with Mary Jo Messenger and Beth Ritsema. *Contemporary Mathematics in Context: A Unified Approach, Course 1 and 3.* Chicago, IL: Everyday Learning, 2003.

_____. *Contemporary Mathematics in Context: A Unified Approach, Course 3.* Chicago, IL: Everyday Learning, 1998.

Coxford, Arthur F. and Zalman P. Usiskin. *Geometry: A Transformation Approach.* River Forest, IL: Laidlaw Brothers, 1975.

Coxford, Arthur F., Zalman Usiskin, Daniel Hirschhorn. *UCSMP Geometry.* Glenview, IL: Scott, Foresman and Co., 1991, 2002.

Devine, Donald F. and Jerome E. Kaufmann. *Mathematics for Elementary Education.* New York: John Wiley & Sons, 1974.

Dressler, Isidore. *Geometry.* New York: AMSCO School Publications, 1973.

_____. *Tenth Year Mathematics: Comprehensive Review Text.* New York: AMSCO School Publications, 1965.

Dupuis, N.F. *Elementary Synthetic Geometry of the Point, Line and Circle in the Plane.* New York: Macmillan and Co., 1894.

Durell, Fletcher and E.E. Arnold. *New Plane Geometry.* New York: Charles E. Merrill Co., 1916.

Education Development Center, Inc. *Connected Geometry.* Chicago, IL: Everyday Learning Co., 2000.

Elander, James E. *Geometry for Decision Making.* Cincinnati, OH: South-Western Publishing Co., 1992.

Fendel, Dan, Diane Resek, Lynne Alper and Sherry Fraser. *Interactive Mathematics Program: Integrated High School Mathematics.* Berkeley, CA: Key Curriculum Press, 1997.

Fischer, Irene and Dustan Hayden. *Geometry.* Boston, MA: Allyn and Bacon, 1965.

Forbes, Jack E and Robert E. Eicholz. *Mathematics for Elementary Teachers.* Reading, MA: Addison-Wesley Publishing Co., 1971.

Foster, Alan G., Jerry J. Cummins and Lee E. Yunker. *Basic Concepts in Geometry: An Introduction to Proof.* Columbus, OH: Charles E. Merrill Publishing Co., 1984.

Gerber, Harvey. *Mathematics For Elementary School Teachers.* Philadelphia, PA: Saunders College Publishing, 1982.

Graham, Malcolm. *Modern Elementary Mathematics.* New York: Harcourt Brace Jovanovich, 1975.

Greenberg, Marvin Jay. *Euclidean and Non-Euclidean Geometries: Development and History.* San Francisco, CA: W. H. Freeman and Co., 1973.

Hart, C. A. and Daniel D. Feldman. *Plane Geometry.* New York: American Book Co., 1911.

Hart, Walter W., Veryl Schult and Henry Swain. *Plane Geometry and Supplements.* Boston, MA: D. C. Heath and Co., 1959.

Herberg, Theodore and Joseph B. Orleans. *A New Geometry for Secondary Schools.* Boston, MA: D. C. Heath and Co., 1958.

Hirsch, Christian R., Mary Ann Norton, Dwight O. Coblentz, Andrew J. Samide and Harold L. Schoen. *Geometry.* 2nd ed. Glenview, IL: Scott, Foresman, 1990.

_____. *Geometry.* Glenview, IL: Scott, Foresman, 1984.

Hobbs, Arles A. *The Elements of Plane Geometry.* New York: A. Lovell & Co., 1896.

Holgate, Thomas F. *Elementary Geometry: Plane and Solid.* New York: Macmillan Co., 1901.

Hopkins, G. Irving. *Inductive Plane Geometry.* Rev. ed. Boston, MA: D. C. Heath & Co., 1902.

Hornbrook, Adelia R. *Concrete Geometry.* New York: American Book Co., 1895.

Jacobs, Harold R. *Geometry.* San Francisco, CA: W. H. Freeman and Co., 1974.

Jacobs, Russell F. and Richard A. Meyer. *Discovering Geometry.* New York: Harcourt Brace Jovanovich, 1972.

Jurgensen, Ray C., and Richard G. Brown. *Basic Geometry.* Boston, MA: Houghton Mifflin Co., 1978.

Jurgensen, Ray C., Richard G. Brown, and John W. Jurgensen. *Geometry.* Evanston, IL: Houghton Mifflin Co., 1990.

_____. *Modern Geometry: Structure and Method.* Boston, MA: Houghton Mifflin Co., 1963.

Jurgensen, Ray C., Alfred J. Donnelly, and Mary P. Dolciani. *Modern School Mathematics: Geometry.* Boston, MA: Houghton Mifflin Co., 1972.

Jurgensen, Ray C., John E. Maier, Alfred J. Donnelly. *Modern Basic Geometry.* Boston, MA: Houghton Mifflin Co., 1973.

Kalin, Robert and Mary Kay Corbitt. *Geometry.* Englewood Cliffs, NJ: Prentice Hall, 1990.

Keedy, Mervin L., Richard E. Jameson, Stanley A. Smith and Eugene Mould. *Exploring Geometry.* New York: Holt, Rinehart and Winston, 1967.

Keenan, Edward P. and Isidore Dressler. *Integrated Mathematics, Course 2.* New York: AMSCO School Publications, 1965, 1981.

Keniston Rachel P. and Jean Tully. *Plane Geometry.* Boston, MA: Ginn and Co., 1946.

Kenner, Morton R., Dwain E. Small and Grace N. Williams. *Concepts of Modern Mathematics, Book 2.* New York: American Book Co., 1963.

Krause, Eugene F. *Mathematics for Elementary Teachers, A Balanced Approach.* Lexington, MA: D. C. Heath and Co., 1987.

Ladd, Norman E. and Paul J. Kelly. *Elements of Geometry.* Glenview, IL: Scott, Foresman, 1965, 1972.

Lang, Serge and Gene Murrow. *Geometry: A High School Course.* New York: Springer-Verlag, 1983.

Lewis, Harry. *Geometry: A Contemporary Course.* Princeton, NJ: D. Van Nostrand Co., 1964.

Lyman, Elmer A. *Plane Geometry.* New York: American Book Co., 1908.

Macnie, John. *Elements of Plane Geometry.* New York: American Book Co., 1895.

Milne, William J. *Plane and Solid Geometry.* New York: American Book Co., 1899.

Moise, Edwin E. *Elementary Geometry From An Advanced Standpoint.* Reading, MA: Addison-Wesley Publishing Co., 1963.

Moise, Edwin E. and Floyd L. Downs, Jr. *Geometry.* 2nd ed. Menlo Park, CA: Addison-Wesley Publishing Co., 1971.

_____. *Geometry.* Reading, MA: Addison-Wesley Publishing Co., 1964.

Morgan, Frank M. and Jane Zartman. *Geometry: Plane, Solid, Coordinate.* Boston, MA: Hougton Mifflin Co., 1963.

Musser, Gary L. and William F. Burger. *Mathematics for Elementary Teachers: A Contemporary Approach.* New York: Macmillan Publishing Co., 1988.

Nichols, Eugene D., Mervine L. Edwards, E. Henry Garland, Sylvia A. Hoffman, Albert Mamary, and William F. Palmer. *Geometry.* Austin, TX: Holt, Rinehart and Winston, 1991.

_____. *Holt Geometry.* New York: Holt, Rinehart and Winston, 1974.

Ohmer, Merlin M. *Elementary Geometry for Teachers.* Reading, MA: Addison-Wesley Publishing Co., 1969.

Palmer, Claude Irwin, Daniel Pomeroy Taylor and Eva Crane Farnum. *Plane Geometry.* 2nd ed. Chicago, IL: Scott, Foresman and Co., 1924.

Palmer, Claude Irwin and Daniel Pemeroy Taylor. *Plane Geometry.* Chicago, IL: Scott, Foresman and Co., 1915.

Pearson, Helen R. and James R. Smart. *Geometry.* Lexington, MA: Ginn and Co., 1971.

Perkins, George. *Elements of Geometry, with Practical Applications.* Hartford, CT: H. H. Hawley and Co., 1849.

Pettee, George D. *Plane Geometry.* New York: Silver Burdett & Co., 1896.

Phillips, Wh. H. H. *Elements of Geometry, and the First Principles of Modern Geometry.* New York: J. W. Schermerhorn and Co., 1874.

Phillips and Fisher. *Plane Geometry.* New York: American Book Company, 1896.

Pilger, Kathy D. *Geometry: for Christian Schools.* Greenville, SC: Bob Jones University Press, 1984.

Posamentier, Alfred S., J. Houston Banks and Robert L. Bannister. *Geometry: Its Elements and Structure.* 2nd ed. New York: McGraw-Hill, 1977.

Quinn, John James. *A Socratic Study of Plane Geometry.* Syracuse, NY: C. W. Bardeen, 1904.

Reichgott, David and Lee R. Spiller. *Today's Geometry.* 3rd ed. New York: Prentice Hall, 1950.

Rhoad, Richard, George Milauskas and Robert Whipple. *Geometry: for Enjoyment and Challenge.* Evanston, IL: McDougal, Littell & Co., 1991.

Ringenberg, Lawrence A. *College Geometry.* New York: John Wiley and Sons, 1968.

Ringenberg, Lawrence A. and Richard S. Presser. *Geometry.* New York: Wiley & Sons, 1971.

Rising, Gerald R., John A. Graham, William T. Baily, Alice M. King and Stephen I. Brown. *Unified Mathematics: Book 2.* Boston, MA: Houghton Mifflin Co., 1985.

Rosskopf, Myron F., Harry Sitomer and George Lenchner. *Geometry.* Morristown, NJ: Silver Burdett Co., 1971.

_____. *Mathematics: Geometry.* Morristown, NJ: Silver Burdett Co., 1966.

Ryan, Patrick J. *Euclidean and Non-Euclidean Geometry: An Analytical Approach.* Cambridge: Cambridge University Press, 1986.

Sanders, Alan. *Elements of Plane and Solid Geometry.* New York: American Book Co., 1903.

Schacht, John F. and Roderick C. McLennan. *Plane Geometry.* New York: Henry Holt and Co., 1957.

School Mathematics Study Group. *Geometry, Student's Text, Part I.* New Haven, CT: Yale University Press, 1961.

Schultze, Arthur and Frank L. Sevenoak. *Plane Geometry.* New York: Macmillan Co., 1925.

Serra, Michael. *Discovering Geometry: An Investigative Approach.* 3rd ed. Emeryville, CA: Key Curriculum Press, 2003, 1997.

_____. *Discovering Geometry: An Inductive Approach.* Emeryville, CA: Key Curriculum Press, 1993.

Seymour, F. Eugene and Paul James Smith. *Plane Geometry.* New York: Macmillan Co., 1941.

Shutts, George C. *Plane and Solid Geometry.* Rev. ed. Boston, MA: Atkinson, Mentzer & Co., 1912.

Sigley, Daniel T. and William T. Stratton. *Plane Geometry.* New York: The Dryden Press, 1948.

Smith, Fred J. *School Geometry.* Chicago, IL: Scott, Foresman & Co., 1897.

Smith, David P. Jr. and Anthony I. Marino. *Plane Geometry.* Berkeley, CA: Charles E. Merrill Co., 1948.

Smith, William Benjamin. *Introductory Modern Geometry of the Point, Ray and Circle.* New York: Macmillan & Co., 1893.

Stubblefield, Beauregard. *An Intuitive Approach to Elementary Geometry.* Belmont, CA: Brooks/Cole Publishing Co., 1969.

Travers, Kenneth J., LeRoy C. Dalton and Katherine P. Layton. *Geometry*. River Forest, IL: Laidlaw Brothers, 1987.

Ulrich, James F. and Joseph N. Payne. *Geometry*. 2nd ed. New York: Harcourt Brace Jovanovich, 1969, 1972.

Usiskin, Zalman, Daniel B. Hirschhorn, Arthur Coxford, Virginia Highstone, Hester Lewellen, Nicholas Oppong, Richard DiBianca and Merilee Maeir. *UCSMP Geometry*. 2nd ed. Glenview, IL: ScottForesman–AddisonWesley 1997; Upper Saddle River, NJ: Prentice Hall, 2002.

Weeks, Arthur W. and Jackson B. Adkins. *A Course in Geometry: Plane and Solid*. Boston, MA: Ginn and Co., 1961.

Weiss, Sol. *Elementary College Mathematics*. Boston, MA: Prindle, Weber & Schmidt, 1977.

_____. *Geometry: Content and Strategy For Teachers*. Tarrytown-on-Hudson, NY: Bodgen & Quigley, 1972.

Welchons, A. M., W. R. Krickenberger and Helen R. Pearson. *Plane Geometry*. Boston, MA: Ginn and Co., 1958.

Welkowitz, Samuel, Harry Sitomer and Daniel W. Snader. *Geometry: Meaning and Mastery*. Philadelphia, PA: The John C. Winston Co., 1950.

Wells, David W., Leroy E. Dalton and Vincent F. Brunner. *Using Geometry*. River Forest, IL: Laidlaw Brothers, 1978.

Wheeler, Ruric E. and Ed R. Wheeler. *Modern Mathematics*. 9th ed. Pacific Grove, CA: Brooks/Cole Publishing Co., 1995.

Wilcox, Marie S. *Geometry: A Modern Approach*. Menlo Park, CA: Addison-Wesley Publishing Co., 1968, 1974.

Young, J. R. *Elements of Geometry, with Notes*. Rev. ed. Philadelphia, PA: Carey, Lea & Blanchard, 1833.